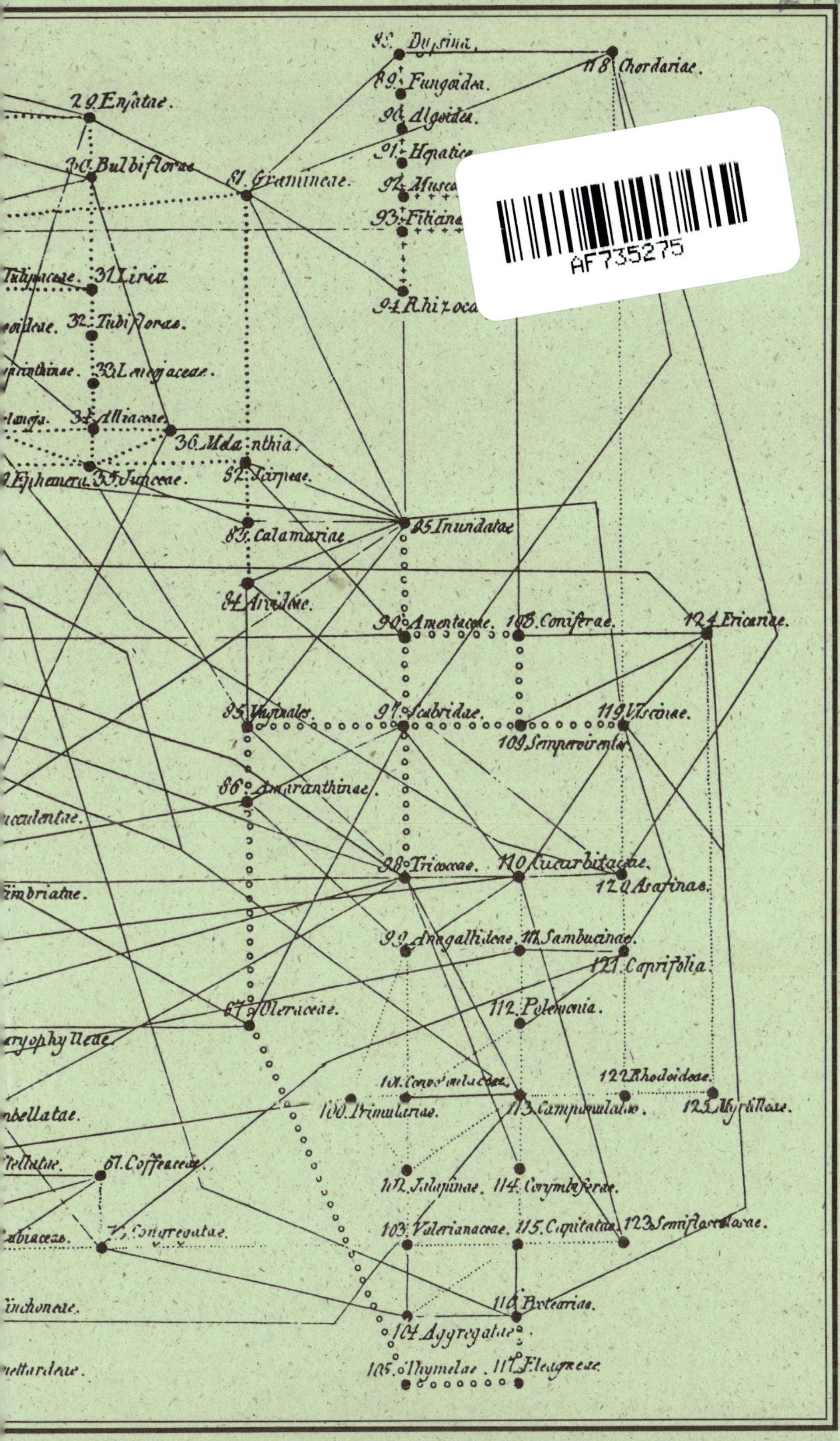
118. Chordariae.
89. Fungoidea.
90. Algoidea.
91. Hepatica
29. Ensatae.
30. Bulbiflorae
81. Gramineae.
93. Filicina
31. Liria.
32. Tubiflorae.
33. Lemnaceae.
94. Rhizoca
34. Alliaceae.
36. Melanthia.
82. Juncae.
35. Junceae.
95. Inundatae.
83. Calamariae.
84. Aroideae.
96. Amentaceae.
108. Coniferae.
124. Ericariae.
85. Vaginales.
97. Scabridae.
119. Viscinae.
109. Sempervirentes.
86. Amaranthinae.
98. Tricoccae.
110. Cucurbitaceae.
120. Asarinae.
99. Anagallideae.
111. Sambucinae.
121. Caprifolia.
87. Oleraceae.
112. Polemonia.
Caryophylleae.
122. Rhodoideae.
101. Convolvulaceae.
100. Primulariae.
113. Campanulatae.
125. Myrtillae.
Umbellatae.
67. Coffeaceae.
102. Jalapinae.
114. Corymbiferae.
Congregatae.
103. Valerianaceae.
115. Capitatae.
123. Semiflosculosae.
Rubiaceae.
116. Proteariae.
Cinchoneae.
104. Aggregatae.
105. Thymelae.
117. Eleagneae.

Interpreting Nature

The distant north, imagined in Amsterdam: the frontispiece to Linnaeus's Flora lapponica *(Amsterdam: Schouten, 1737)*

Interpreting Nature

The Science of Living Form from Linnaeus to Kant

James L. Larson

The Johns Hopkins University Press
Baltimore and London

Printed in the United States of America on acid-free paper
03 02 01 00 99 98 97 96 95 94 5 4 3 2 1

The Johns Hopkins University Press
2715 North Charles Street
Baltimore, Maryland 21218-4319
The Johns Hopkins Press Ltd., London

ISBN 0-8018-4840-7

Library of Congress Cataloging-in-Publication Data will be found at the end of this book.

A catalog record for this book is available from the British Library.

Endpapers. Front: *Table of affinities for the vegetable kingdom from A.G.J.C. Batsch*, Tabula affinitatum regni vegetabilis *(Jena: Landes-Industrie Comptoir, 1802)*. Back: *Table of affinities for the higher animals conceived by Johann Hermann*, Tabula affinitatum animalium *(Strasbourg: Treuttel, 1783)*.

Contents

List of Illustrations

Acknowledgments

This list of debts properly begins with the names of two libraries. The Kofoid Collection in the Biology Library at the University of California, Berkeley, has been the center of my research for more years than I care to count. I have also relied upon the resources of Carolina Redeviva in Uppsala, Sweden. The efficiency, helpfulness, and intelligence of the staffs at these two institutions are beyond praise.

It is a pleasure to thank Gunnar Broberg, Phillip Sloan, Peter Stevens, and Shirley Roe for photocopies of otherwise unobtainable texts and for their efforts to keep me au courant with their specialties. Art Quinn, Phillip Sloan, Peter Stevens, and Mary Winsor offerred valuable suggestions on the manuscript, many of which are reflected in this book. I am grateful to these colleagues for taking the time to analyze an unpublished work so meticulously in the midst of the school year.

Earlier versions of two portions of this book have appeared in the *Journal of the History of Biology* ("Not Without a Plan: Geography and Natural History in the Late Eighteenth Century," *JHB* 19 [1986: 447–88]) and in *The Quantifying Spirit in the Eighteenth Century*, edited by Tore Frängsmyr, John Heilbron, and Robin Rider (Berkeley: University of California Press, 1990). I thank D. Reidel Publishing Company of Dordrecht and the University of California Press for permission to adapt this material.

Finally, I thank the staff of the Office for the History of Science and Technology at the University of California, Berkeley, most especially Diana Wear, for coming to the rescue and preparing the final typescript of this book with celerity and vigilance.

Interpreting Nature

Introduction

My title echoes Diderot's *Pensées sur l'interpretation de la nature*, and like Diderot's little book, these studies deal with the attempt by Europeans of the eighteenth century to conceive living nature in purely physical terms, as a unified and self-contained system. Diderot approached this scientific transformation as an observer, and he was concerned above all to offer counsel and encouragement to French participants. Two and a half centuries after the event, my vantage is that of the historian. My account is therefore only an analytical construct, without the sense for potential development and social consequences that Diderot took for granted. What these studies offer instead is a perspective, precisely the perspective that allows us to give special importance to this instauration in eighteenth-century natural history and physiology.

From almost the very beginning, with the publication of the first volumes of Buffon's *Histoire naturelle* in 1749 and the *Encyclopédie* in 1751, this particular scientific transformation involved not only naturalists, physicians, and physiologists but also theologians, philosophers, and publicists, who saw to it that every step of the way was accompanied by commentary of all kinds. The principal stages of the transformation have since been studied endlessly, and the literature on individuals, methods, concepts, and controversies is not only very large but resists summary.

My own debt to the secondary literature is spelled out in the following studies. Here, I want to indicate in a more general way some of the persistent problems one confronts in working with this body of commentary. Considering the amount of secondary literature, it comes as something of a surprise to discover so many gaps in the historical record: information about the problems that provoked a conceptual reform is spotty, there are only a few institutional studies of the life sciences, and our knowledge of the sciences of natural history and physiology outside the world of French scholarship is intermittently reliable and far from coherent.

Not only are there holes in the historical record; there are many failures to understand the past on its own terms. Our feel for the ways in which naturalists and physiologists posed their questions is often determined by the commentator's position in history. Archaic reasoning is corrected or suppressed. Useful methods and results are emphasized, errors explained away. In fact, much of the research on the life sciences in the eighteenth and early nineteenth centuries is only a translation from a language grown partially incomprehensible into our own thoroughly familiar conceptual terms.

Let me offer a simple example from Ernst Mayr's *The Growth of Biological Thought.* In a section on the origins of evolutionary thought in Germany, Mayr touched upon the German reaction to the mechanist conception of nature in western Europe.

> To reduce every phenomenon and process in the organic world to movements and forces, or to heat and gravity, was, quite rightly, unacceptable to most naturalists, who posited numerous alternatives. . . . Those who did not want to invoke God to explain everything in nature elaborated a new view of nature, strongly influenced by Leibniz, which stressed quality, development, uniqueness, and usually a finalist component.[1]

Professor Mayr explained the enthusiastic reception of this view as part of the prevailing disgust with mechanist explanations.

His discussion echoes an older German historical tradition, which viewed the last quarter of the eighteenth century as a shift from mechanist to organic patterns of thought. Ernst Troeltsch, whose work did so much to shape perception of the importance of this shift, summed up the tradition succinctly in *Der deutsche Idealismus.* French and English culture were characterized by a mechanistic and atomistic conception of nature, a conception that dominated all areas of thought. German culture, on the other hand, was characterized by an organic and dynamic conception that made nature subservient to the spiritual aims of the universe.[2] These statements by Troeltsch and Mayr, separated by forty years, are part of a grand historical scheme—the rise of a specifically German way of thought on the basis of a common European tradition but in contrast to the French and English enlightenment.

In the English-speaking world, a rather different view of the eighteenth-century life sciences has prevailed. Interpretations of preevolutionary periods are often anchored in the Darwinian synthesis, and the great contributions of thinkers and scientists are reduced to programmatic statements introducing ideas of history and descent into atemporal sciences. To avoid offense, let me simply mention an older but often-cited book in the history of ideas, *Forerunners of Darwin*, which in its very title can be seen as thoroughly typical of this tradition.[3]

The slogans of these interpretations, from *mechanism* to *organism* and from *natural history* to the *history of nature*, indicate two of the principal directions the history of the life sciences in the eighteenth century has taken. Neither tradition is much concerned with the actual problems confronting the life sciences between 1740 and 1790; arguments based on organic conceptions are oriented toward the period after 1802; arguments for the historization of the life sciences look forward to the Darwinian synthesis. Both traditions concentrate upon programmatic statements by the great personalities, as Friedrich Meinecke once called them,[4] and in this, the traditions are representative of the tendency to make principle or theory or worldview the driving force in the growth of scientific ideas.

Recently, another large body of commentary has built up around the origin of the term *biology* early in the nineteenth century. Although older historians have often touched on the subject,[5] my discussion takes Michel Foucault's *Les mots et les choses* as its point of departure. In dealing with the epistemological lattice of natural history, Foucault remarked that a historian must not use the term *biology* in speaking of eighteenth-century science.

> We have sought to write histories of biology in the eighteenth century; but this does not take into account that biology did not exist and that a division of knowledge familiar to us for more than fifty years cannot apply to an earlier period. And if biology was unknown, there was a very simple reason, namely, that life itself did not exist. There were only living beings who appeared through the epistemological lattice of natural history.[6]

Whatever the value of Foucault's observation as historiography, it has had the merit of awakening historians from their dogmatic slumbers. Few question the importance of the discontinuity Foucault established between natural history and biology, but most specialists object to his conflation of comparative anatomy and biology. Joseph Schiller said that Foucault's views were unsupported by fact, and he published an analysis of the concept of organization to show that the emergence of the term *biology* was not an instance of spontaneous generation.[7] Schiller supported Lamarck as an alternative source for the new conception of life. Claire Salomon-Bayet proposed institutional experimentation on living beings as an alternative source for the new science.[8] Several analogous proposals have been made, all analyzed and summed up in a case study by Joseph A. Caron, who argued that the emergence of the term *biology* was accompanied by complex historical, social, and institutional changes and cannot be reduced to single disciplines and historic figures.[9]

Among German historians of science, parallel claims on behalf of Gottfried Reinhold Treviranus have been made several times in recent years. Of these, Birgitte Hoppe offered a good introduction to the elements of Treviranus's *Biologie*.[10] Wolf Lepenies described Treviranus as part of the transition from natural history to the history of nature and argued that Treviranus's use of the spiral as metaphor for developmental processes was an important step in the new orientation toward history.[11] Timothy Lenoir was more cautious. He admitted that Treviranus "attempted to found a new science" but concluded that "having stated their vision of biology as a unified program for integrating the organic world, the 'biologists' were forced just as quickly to retreat from it."[12]

From the vantage of a historian interested primarily in eighteenth-century science, commentary on the origins of biology shares at least one fault with the older traditions described above. Discussion is largely oriented toward the nineteenth century, while eighteenth-century science is used as a quarry. Judged by the standards of the period they profess to investigate, the main lines of inquiry into the history of the life sciences of the later eighteenth century are inadequate. Historians of that period, men as different as Herder and Justus Möser, were attempting after all to value a period in terms of its unique content and intrinsic meaning, and they interested themselves in far more than the alluring but abstract possibilities of programmatic statements. Our best hope of discovering the integrity of the life sciences in that period lies in following their example in studying the social contexts, working methods, and actual problems that professional scientists proposed to solve.

Historians have seldom considered the period following 1740 as a unit, although the life sciences underwent important changes in content and method, changes whose effects can be traced across the entire range of scientific investigation. The theological conception of an immutable natural order was replaced by a secular view of nature as an autonomous power. The time scale against which nature pursued the continuities of her own development increased throughout the period and even began to include discontinuities, or "revolutions." Collections of objects, if not their arrangement, ceased to be Eurocentric, thanks to the efforts of naturalists, patrons, government companies, and government expeditions. Botanists worked out a natural method for plants. Zoologists laid down the main lines for the development of comparative anatomy. Physiologists ran up against the limits of mechanical medicine and changed course, uniting physicomechanical and teleological principles in their explanations.

There are serious obstacles to viewing these disparate developments as elements of a coherent pattern, obstacles imposed by the organization of scientific investigation between 1740 and 1790. The life sciences—natural history and physiology—were divided by alternative approaches and resulted in different kinds of intelligibility. It is far from easy to describe the persons, the "naturalists" and the "physiologists," who studied living forms. This difficulty involves a related problem, the base of support for scientific investigation. Let me outline each of these problems in a little detail.

When Treviranus defined biology in 1802, he was concerned above all with an ambiguity in all previous study of living forms. The sciences of natural history, never properly limited or defined, approached living forms as identifiable kinds. However modified and refined, the procedures of these sciences had remained basically the same: the specific kinds offered a point of departure for a well-defined problematic, which placed each kind against an atemporal logical horizon, simultaneously containing and situating it—"mere registers of names," wrote Treviranus. An alternative approach, which studied the "conditions, laws, and causes of life," belonged essentially to medicine, physiology, and pathology.[13]

The contrast in epistemological intention that separated the qualitative, observational, and descriptive enterprise of the naturalist from the analytic and often experimental enterprise of the physiologist was obvious. The two approaches resulted in two kinds of intelligibility: the one, Aristotelian, of classification; the other, mechanist or vitalist, of causality. Treviranus's new science was supposed to bring the perspectives of natural history and physiology together and to read the forms and appearances of life over against the conditions, laws, and causes of life. The ambition was beyond Treviranus's grasp in 1802; it was well beyond the imagination of all but a few naturalists and physiologists in the eighteenth century.

It is far less easy to describe the naturalists and physiologists who investigated living form in the eighteenth century. Jacques Roger has pointed out that there are few sociological and institutional studies of the life sciences.[14] In the absence of these studies, it is almost impossible to discuss the place of naturalists and physiologists in scientific institutions, the ways in which they worked, and the kinds of science they created.

Obviously, medical doctors dominated the study of living beings, because they alone received regular instruction in these matters. Anatomy, physiology, and materia medica were parts of medical curricula, and a professorship in medicine was one way in which a living could be extracted from these studies. In countries where universities remained centers of intellectual life, like the Netherlands, the Reich, and Scandinavia, this state of affairs persisted throughout the century, although ancillary research constituted an increasingly large part of professorial activity. "Hermann Boerhaave and Friedrich Hoffmann," observed Roger, "were primarily professors of medicine, even if their teaching included a physiology of their own. Carl Linnaeus and Albrecht von Haller were also professors of medicine, but the former was first of all a botanist and the latter an experimental physiologist."[15] Increasing specialization mirrored the transformation of the sciences themselves, from ancillary studies to independent disciplines.

After 1750, opportunities for research careers in natural history and physiology expanded as more academies were founded.[16] The academician was characterized by his commitment to research, just as professors supposedly dedicated themselves to teaching. In practice, the distinction was not so neat, just as relations between universities and academies were complex and various. St. Petersburg combined the academy with instruction; Hannoverian officials regarded the academy and the university at Göttingen as complementary; and the Elector of Bavaria established his academy over the protests of professors and excluded them as members. Two general results emerge from this situation: after 1750, as the publication of academic transactions and commentaries multiplied, research in natural history and physiology increased; thanks to the academic insistence upon disciplinary competence, the quality of research improved.

Academicians and professors were not, however, the only important contributors to natural history and physiology in the eighteenth century. Amateurs played an important role in eighteenth-century science—and not just as patrons. The social dimensions of this situation and its influence on the practice of science have been studied only in such cases as Charles Bonnet, John Turberville Needham, and Abraham Trembley. "What these amateurs had in common was the fact that they were self-taught and did not belong to the traditional medical profession. They had learned natural history through reading and observation, but this did not prevent them from making some of the most striking discoveries of the century."[17] A related factor is the public enthusiasm for scientific curiosities, a subject that has its frivolous aspects, but without which a distinct scientific ethos could not have emerged during the final decades of the ancien régime.

In central and northern Europe, the rationalizing bureaucracies of the absolutist states did the most to promote this ethos. The will of the state to dominate, organize, and exploit human and material resources touched many aspects of life in these regions, including the training and protection of technicians and scientists of all kinds. Officials found employment for naturalists and physiologists not only at academies and universities but also in medical and agricultural collegia, on army staffs, and in admiralty offices.[18] These general factors—ambiguities in the study of vital phenomena and the lack of any comprehensive analysis of either the persons who studied living beings or their place in scientific institutions and in society at large—determined the scope and the limitations of the following studies.

Since the entire range of natural history and physiology would have required an encyclopedia, I have limited myself to the sciences of living form. Practically indispensable, the limitation is not merely conventional. Living forms—their production, growth, organization, modifications, characters, affinities, and distributions—were the central concern of natural history and physiology throughout the century. The sciences of living form included two branches of taxonomy: the primary divisions of the plant and animal kingdoms and the study of species and their modifications. A separate but related inquiry studied the organization of the living world as a whole, the distribution of natural forms. These sciences, separate divisions of natural history, took the *Gestalt* of living forms as their point of departure. The science of formation, *Bildung*, was a part of physiology and studied the principles and mechanisms of organic generation. Naturalists and physiologists treated these inquiries as largely independent. They are still best studied separately.

My second limitation was geographical. I concentrated on research originally published in the Netherlands, the Reich, Russia, and Scandinavia not only because this is the area of my own competence but also because science in northern and central Europe has been comparatively neglected. No specialist in eighteenth-century science can ignore the Académie Royale des Sciences in Paris, however, and I have not hesitated to open or close individual studies with work by Parisians that determined the direction of research elsewhere.

Because organized learning implies an institutional framework, these studies are limited to the opportunities and constraints imposed by academies and universities, and I have taken John Heilbron's work on early modern physics as a model. "The belief common among historians, who concern themselves only with Britain and France, that university professors made only a small and continuously declining contribution to natural philosophy during the seventeenth and eighteenth centuries also falls before the facts."[19]

Finally, there is the limitation of time. Conceptual reform in the sciences is never a product of one day, one year, or even one generation. In the case of the reform program of the eighteenth-century life sciences, the significant changes were concentrated in the second half of the century, when pertinent shifts in method and content occurred in 1760 and 1790. This book limits itself to two periods, 1740–60 and 1760–90, but includes relevant material from earlier and later periods.

The limitations of subject matter, time, place, and persons are intended to point up what I take to be characteristic of the life sciences between 1740 and 1790. During the first period, 1740–60, naturalists and physiologists argued for the study of physical relations, secondary causes, and intelligible principles; they argued against methods based on superficial analogies, first causes, and divine intentions. Obviously, the program involved questions about the relative autonomy of living nature, questions that laymen of the time regarded as uncomfortable and undesirable for any number of academic, social, and theological reasons. The program was difficult to defend, and individual scientists compromised in different but characteristic ways between physical and metaphysical principles.

During the second period, 1760–90, naturalists and physiologists defended the secular program of the scientific enlightenment with a tenacity and rigor not found in the earlier period. As individual disciplines in natural history and physiology established the autonomy of principles and processes proper to living forms, scientists began the long process of rooting out or modifying assumptions and concepts inherited from the older theological conception of natural order. Gradually, they came to view the whole of living nature as a dynamic and self-regulating economy. But this economy was not yet thought to unfold in a dialectical process or along an irreversible time line. Ideas concerning the historical development of nature were loosely attached to the disciplines of exact observation and did not take root in the sciences of living form in the eighteenth century.

I

Two Generations

A Plutarchian comparison of Linnaeus and Buffon was often made in the later eighteenth century.[1] Linnaeus, who embodied the systematic spirit, had inspired much of the academic natural history written in northern and central Europe. Buffon's *Histoire naturelle*, comprehending almost the entire range of the life sciences, was regarded at first as a suspect product of French materialism.[2] The work was an immediate popular success, but Buffon's penchant for abstract speculation limited scientific recognition, and it was only after an extended campaign that Buffon's unorthodox ideas began to receive serious attention from naturalists and physiologists. The struggle for and against Linnaeus and Buffon transformed the two men into symbolic figures.

Linnaeus, many believed, had introduced order and precision into natural history and had worked with uncommon zeal to extend scientific knowledge of organized beings. Buffon, on the other hand, recognized neither order, plan, nor liaison in sublunary nature and replaced method with fantastic and unprovable hypotheses about the autonomy of matter, the generation of living forms, and physical truth.

The partisans of Buffon, needless to say, had very different ideas. Linnaeus and the Linnaeans were practitioners of a reactionary science, were obsessed with nomenclature and definitions, and had substituted arbitrary systems for the study of nature. Buffon, on the other hand, sustained by the laborious investigations of Louis-Jean-Marie Daubenton, had extended the limits of science with his philosophical spirit, while the charm of his eloquence reached far beyond lecture halls and laboratories and won science friends among the great and among men of the world.

Those who experienced the renewal of natural history at firsthand were convinced that Linnaeus and Buffon had given a new thrust to human knowledge. "The one and the other," wrote Peter Simon Pallas, "happily for natural history, have appeared in the same century to advance that science toward perfection with giant steps, although by different, and even opposed paths." Pallas considered these paths so different, that he counseled students of natural history to choose between them. "Natural history, in this respect, has experienced the fate of other sciences. Students studying today must choose between these two great masters."[3] But Pallas exaggerated. His own career shows that a proper course could not be charted from the positions of Linnaeus or Buffon alone; the path of reason led between their opposed positions. And there were other stars in the sky. Admittedly, not all of these stars were of the same magnitude, but there is at least one more that requires attention.

Albrecht von Haller was a name well known to both Linnaeus and Buffon. The best sign of his scientific importance is that both men feared him and dreaded his criticism. And they treated him as an equal, a courtesy they rarely extended to other mortals. Haller's name is often omitted from accounts of eighteenth-century natural history, because his great contributions were to the science of physiology, a very different science with a very different audience. But Haller was also a major player in natural history, where his erudition, industry, and aggression won him international authority.

This triumvirate, George Louis Leclerc, Comte de Buffon (1707–88), Albrecht von Haller (1708–77), and Carl von Linné, or Linnaeus (1707–78), dominated the study of the sciences of living form on the European continent at midcentury, an influence they continued to exercise until the end of their lives and even beyond. Between 1734 and 1741 they had found academic niches in Paris, Göttingen, and Uppsala, and in short order their work brought them international recognition. Their first encounter with one another on the great stage of Europe was so unpleasant that the three thereafter avoided contact and involved others in their quarrels. In mitigation, it must be pointed out that their very different educations, cultures, and conceptions of science did not leave much room for collaboration. But their first encounter is illuminating not only of the contenders' personalities but of their sciences.

After many years of study, Buffon published the first three volumes of the *Histoire naturelle* in 1749. In these volumes, Buffon confronted older natural history with a new analytical and critical spirit and concerned himself with different conceptions of procedure, evidence, and truth. Historians of science have described his program as a decisive turning point in the life sciences, comparable in some ways to the great innovations of the preceding century.[4] Certainly the pronounced secularity of Buffon's program produced many points of friction with the science taught and practiced elsewhere in Europe.

The charter of the new method was Buffon's preliminary discourse, "De

The triumvirate: Buffon, Haller, and Linnaeus

la manière d'étudier et de traiter l'histoire naturelle." Buffon proposed to consider natural order in relation to the human capacity for knowledge and to liberate natural history from its preoccupation with classification. He seemed to argue that systems were bad in themselves and negligible in relation to everything else, an argument that working naturalists received as a rejection of science itself.[5]

Buffon's discussion opened with a commonplace: a student confronted with the great number of natural objects had two options: the first was to have no method; the second was to relate everything to one particular system. Systems abridged labor, aided memory, and offered an interesting sequence of ideas, but they also led men to judge wholes on the basis of parts, to assemble arbitrary groups, and to divide nature at points where she was indivisible. When applied dogmatically, a system insinuated a false structure of relations into the order of nature, producing a spurious and straitened regularity.

As a prime example of this benighted science, Buffon took the Linnaean system. Certainly there were equally appropriate works by Buffon's colleagues in the Paris academy, but he singled out Linnaeus for a hostile and aggressive attack.[6] The Linnaean sexual system in botany resulted in a bizarre assemblage, placing entirely unlike plants in the same class. The orders of quadrupeds were equally arbitrary and violated likenesses and differences that might have justified other divisions.[7] Buffon brusquely rejected Linnaean methods and with them a long tradition of inflated claims and barren scholasticism:

> First causes are forever hidden from us; the general results of these causes are as difficult for us to know as the causes themselves; all that is possible for us is to apperceive some particular effects, to compare them, to combine them, and finally to recognize an order relative to our own nature rather than convenable to the existence of the things that we consider.[8]

"An order relative to our own nature" is the operative phrase in this statement. Buffon's conception of relative truth is central to any understanding of his thought on system and method. It provides a key not only to his belief in an intelligible natural order but also to the alternative approach to natural history advocated in the preliminary discourse.[9] An understanding of Buffon's idea of physical or probable truth is at the very center of the controversies examined in these studies.

A naturalist's ideas about living forms necessarily differed from their reality, but the arrangement of these ideas, Buffon argued, was not therefore arbitrary. Relations among ideas were true or false as they corresponded or not to relations among the objects signified. When relations among ideas were invariable and always the same relative to himself, a naturalist did not need to doubt their truth, for truth consisted in the connexity of signs corresponding to the real and necessary connexity of objects.[10] By establishing parallel orders between ideas and events, Buffon founded scientific generalizations upon real—formal and successive—relations among objects:

> Physical truths are not at all arbitrary and do not depend upon us; rather than being founded on suppositions we have made, they rest upon facts; a sequence of like facts or, if you prefer, a frequent repetition and uninterrupted succession of the same events constitute the essence of physical truth; what we call physical truth, then, is only a probability, but a probability so great that it is equivalent to certainty. (ibid., p. 24)

This theory enabled Buffon to study the constant and general order of phenomena but without falling into strict phenomenalism and without advancing dogmatic claims for his results.

On this basis, Buffon proposed a convention alternative to Linnaeus's use of hierarchical divisions. Buffon wished to begin by studying natural forms that interest us because of their relation to us. Necessary or useful forms come first, because we know best what is familiar. Admittedly, the order of familiarity was subject to the objection against all artificial methods: it assembled heterogeneous objects. But method existed for purposes of study and considered things in the order in which they were found ordinarily. In this framework, the best way to advance science was to give a complete description and an exact history of each thing in particular. The study of nature would limit itself at first to these descriptions.

Later, a naturalist would generalize facts, connect them by means of analogy, and exhibit the dependence of particular effects upon those more general. The procedure guided thought, sustained natural order, and clarified and extended the view—and all without misleading the observer. The most important element in the study of nature was to recognize what was real in the subject without adulterating it.

Buffon was quite definite about what was real in his subject. Linnaean class concepts were unreal because they lacked physical content; the projected unity had no objective correlative. To be meaningful, then, a class concept had to indicate a physical identity with determinate physical effects. The model for this identity was furnished by a group of individuals of the same species. In the preliminary discourse, species were said to be "real" or "natural" but to be grounded somewhat indecisively upon relations of both similarity and succession. By the time of publication, though, Buffon had convinced himself that, in the matter of species criteria, reproduction took precedence over the formal likenesses of contemporary systematics. "The species," he wrote in 1753, "is nothing more than a constant succession of like individuals who reproduce one another" (ibid., p. 356). The reality of the species consisted precisely in the "frequent repetition and uninterrupted succession of the same event," which constituted the essence of physical truth.

The reality of species implied a similar reality in more inclusive groups of natural forms, a conviction reinforced by the results of Daubenton's anatomies. Daubenton's uniform descriptions revealed concordances among quadrupeds that suggested the idea of "a first design, upon which everything seems to have been conceived."

> If, in the immense variety presented by all the animals that populate the universe, we choose one animal . . . as a basis for our knowledge and relate to it, by means of comparison, other organized beings, we shall find that, although all these beings exist solitarily and . . . vary by degrees graduated to infinity, there exists at the same time a primitive and general design that one can follow distantly, whose degradations are rather more gradual than those of figures and other apparent relations. . . . In the very parts that contribute most to the variety of the exterior form, there is a prodigious resemblance that necessarily recalls to us the idea of a first design, on whose basis everything seems to have been conceived. (ibid., pp. 353–54)

In the same passage, Buffon questioned whether neighboring forms, too distant to allow crossing, might not descend from a common branch. The question led to a related problem, the meaning of the general term *family*. Naturalists who established families of animals and plants did not seem to recognize the consequences.

> If these families exist, in effect, they have only been able to form themselves by mixing, successive variation, and degeneration of the originary species. . . . Each family, both animal and vegetable, has a single branch, and . . . all animals have come from a single animal, which in the succession of time has produced, by perfecting itself and by degeneration, all the races of other animals. (ibid., pp. 354–55)

If class concepts have a meaning—that is, if they have a basis in physical reality—they must signify community of origin. Buffon's insight, understood in all its extent, would interpret the unity of plan in animals as a sign of unity of descent, of common heredity.

As Daubenton extended the results of his anatomies, Buffon returned to "that plan always the same," which when grasped by the mind "seemed a faithful exemplar of living nature, both the simplest and the most general view under which we may consider her" (ibid., p. 389). Buffon's general meaning is clear. He regarded the unity of plan found in the animal series as a consequence and sign of descent, of common heredity, an idea directly linked to his insistence upon an objective correlative for class concepts, that is, a real physical identity with real physical effects. By giving to systematic distribution a genetic significance, Buffon redefined class concepts as expressions of a real process by which groups were constituted in nature.

On many points, Buffon's thought was obscure. In his early volumes, he said little about the composition of the groups whose unity he affirmed, their relations with other groups, or the process of their formation. He offered his new point of view as a subject for speculation, not as finished doctrine. He made use of traditional terms in his presentation—*continuity*, for example, and *essential parts*—that did not represent his thought adequately. And although he considered class concepts intelligible only to the degree that they rested upon a physical identity with physical effects, he continued to represent those effects in static and formal language. His skepticism as to the value of an abstract concept was accompanied by the assumption that scientific concepts represent common elements in a collection of similar or comparable forms.

Buffon's obscurity and equivocation hid from his contemporaries the boldness and independence of his fundamental insight. What was more obvious to the learned world was the superficiality of Buffon's discussion. The self-assurance of the writer and the arrogance of the gentleman could not hide the limitations of Buffon's knowledge not only of the scientific data but of the questions posed by naturalists. His antipathy to Linnaeus was in part personal, the expression of a total opposition of temperament, vocation, and ambition. More importantly, though, Buffon rejected the arid nomenclature and monotonous characters of Linnaean formalism, which, he said, made the language of science more difficult than science itself (ibid., p. 12). He saw Linnaean science as a survival of scholasticism, and he made Linnaeus the symbol of this rebarbative science. Members of the older generation in the Paris academy protested, warning Buffon that he had not grasped the spirit of Linnaean science, but in vain.[11]

In short order, a German translation of the popular *Histoire naturelle* was announced. Within the year, the first volume appeared under the imprimatur of the Elector of Saxony, with a preface by Albrecht von Haller and translation and commentary by Abraham Kästner.[12] Here, if anywhere, we expect to find an authoritative reaction by the scientific establishment to Buffon's bold new ideas.

Haller began by recalling the vogue for rational natural philosophy in the preceding century, a vogue that quickly led to controversy, after which oblivion swallowed the victors and the vanquished. Simultaneously, craftsmen perfected instruments to examine nature, "telescopes, rounder glass drops, more precise divisions of measurement, syringes, and scalpels," inventions "that did more for the enlargement of science than the imaginative mind of Descartes, than of the father of classification, Aristotle, and than the erudite Gassendi."[13] Mathematical learning spread over Europe and taught men to creep where once they had flown. And every step that brought science closer to nature showed how little she resembled the pictures drawn by learned men.

This realization led to skepticism and the distrust of hypotheses, an attitude that stifled the search for truth. Human knowledge consisted of fragments, whose parts lacked coherence. If men did not supplement certainty with probability and conjecture, science would become impossible. Hypotheses were not truth, but they led to it.

Linnaeus, for example, had based his method according to stamens and pistils on the hypothesis of sexuality in plants, that is, an analogy with the sexual parts of animals. Although Linnaeus's principles were only probable, they had performed a great service. Botanists were stimulated to describe the flowers and fruits of all plants. "Botany raised her head above all the other sciences; she was not only next to perfection . . . but she communicated her laws to the whole kingdom of nature. Zoologists and physicians received their laws from her, as the Romans from Areopagus, and submitted to them" (ibid., p. xvi).

Hypotheses, including the arbitrary classifications of botany, stimulated questions whose answers were to be sought in nature. Some questions would lead to new discoveries; others—including Buffon's improbable assertions—might lead to shipwreck. In Buffon one found learning, experience, and insight, but, Haller added, this was an author who always pressed beyond what he knew, and Haller advised the reader to exercise caution in reading Buffon (ibid., p. xxii).

Everything in the first volume of the German translation bears the signs of a hastily conceived and loosely coordinated editorial policy. Both Haller and Kästner were busy men, publishing widely on many subjects, and inquired not at all closely into the basis for Buffon's criticism of systematics. Haller's argument—that hypotheses are questions addressed to nature—was an early version of an argument made by Kant in the first introduction to the *Kritik der Urtheilskraft*. The argument, important as it is, is tangential to Buffon's distinction between "abstract" and "physical" systems.

Kästner involved himself in a fundamental mistake from the first note in his commentary. He read Buffon's distinction between abstract and physical truth against the ambitions of contemporary systematics and interpreted the distinction as the commonplace separation of artificial and natural methods. His reading did not quite fit, and Kästner was actively exercised by the concluding discussion of different kinds of truth in Buffon's "Premier discours."[14] It is not surprising that Kästner ignored Buffon's discussion of the species as models of physical continuity: Kästner did not recognize what was at stake.

This interpretation of Buffon's position by two authorities set a pattern for the reception of the German translation. As Phillip Sloan has shown, reviews concentrated on Buffon's cosmological, geological, and generational theories. When reviewers took note of Buffon's ideas on classification, they treated them, as had Haller and Kästner, as "groundless and offensive judgments."[15]

A second volume of the German translation was scheduled for publication in 1752. During 1750 or 1751, Haller was forced to confront Buffon's revival of epigenetic theory, a theory not so very different from Haller's own at the time. Haller had been arguing that animals were generated successively from a fluid in accordance with general laws; he was even toying with the force of attraction as an organizing mechanism.[16] In Buffon's treatise on generation, Haller confronted the implications of his own position. His preface not only rejected those implications but laid down the lines of Haller's subsequent development as the champion of preformation. In the mind of Buffon, this could only have identified Haller with the generation of Réaumur. Everything condemned the two men to oppose one another.

But what of Linnaeus, the original subject of their confrontation? Linnaeus did not read modern languages and can have followed this controversy only at secondhand. Clearly, the story lost nothing in the telling. Buffon's treatment of Linnaean method made Linnaeus Buffon's lifelong enemy. Thereafter, Linnaeus carefully recorded every triumph over his foe. In 1762 Linnaeus was elected one of the six foreign members of the Paris academy, an honor he regarded as "the high point of my career," not least because he regarded the Paris academy as the citadel of Buffonian science.[17] In 1774, Linnaeus celebrated an even greater coup. The botanical garden in Paris was rearranged according to the Linnaean system. Linnaeus recorded the event for posterity: "Buffon, who always wrote against me and lived in the botanical garden in Paris as intendant, must now range the plants after my system, *nolens volens*."[18]

As for Haller, Linnaeus had learned not to expect much from that quarter. They had quarreled in 1736, when Linnaeus contradicted Haller in the *Flora suecia*. Thereafter, Haller had attacked Linnaeus publicly, chiefly on matters of nomenclature. Linnaeus, said Haller, "regards himself as another Adam and has named all the animals according to their characteristics, without regard to his predecessors." Linnaeus replied in kind. "Haller," he said, "regards himself as infallible as the pope." We do not know whether Linnaeus heard of Haller's praise in the Buffon preface, but his pleasure would have been poisoned by the remark that the sexuality of plants was hypothetical and that the famed sexual system was arbitrary. Haller and Linnaeus, said Abraham Bäck, resembled Caesar and Pompey. "The one, our Linnaeus, could not tolerate anyone equally powerful, and the other, Herr von Haller, no one mightier, or the reverse."[19]

The triumvirs' first encounter shows, if nothing else, that the traditional emphasis on differences in the aims and methods of the three men has not been misplaced. The essentialism of Linnaeus and the Linnaeans differed as much from Buffon's synthesis of materialism, attractionism, and the philosophy of Christian Wolff as this in turn differed from Haller's complex mix of pietism, mechanism, and experimentalism. But we should not exaggerate the importance of this confrontation in the science of the antagonists. The incident created lasting animosities, but no one changed course on this basis, and the event did not touch their scientific reputations. In 1750 Buffon, Haller, and Linnaeus were at the height of their careers. In the next decade they published the works that have secured their scientific immortality.[20]

Nor should the differences in their conceptual worlds obscure the similarities in some of the assumptions shared by the three men. One set of assumptions can be detected in common conceptual elements that recur in their work—the notions of continuity, purpose, and plan, for example, as well as the fixity of the species and the preformation of the organism in a germ, in the medulla, or in an internal mold. These were products of very diverse origin—Christian scripture, Aristotelian philosophy, and mechanist "physicks," to name only the most obvious. Taken in combination, as they had been since the late seventeenth century, they were components of an immutable natural order, a system of regular, constant, and universal relations among created beings, ultimately founded in the will of God.

As naturalists and physiologists, however, the triumvirs could not help responding to conceptions of living nature as dynamic and relatively free from divine intervention. Investigation could in good conscience put an end to the traffic in final causes and divine intentions and give itself over to the study of physical forms and processes. The natural history and physiology of Buffon, Haller, and Linnaeus attempted in very different and characteristic ways to balance the claims of these alternative positions.

After 1760 the triumvirs began to experience a new and very different kind of criticism. A younger generation subjected their cherished concepts and observations to close scrutiny and rejected them out of hand. The criticism was hard to answer, because it was based on precise definitions and close observations. For the most part, the new criticism concerned specialized questions, but it could not be disregarded on that account. Very often, these questions formed the keystones of larger theoretical constructions. What made the situation even harder to bear was the disrespectful attitude of the younger men toward their elders and betters.

Shortly after Albrecht von Haller published *Sur la formation du coeur dans le poulet* (1758), in which he demonstrated the truth of preformation, he received a dissertation from Berlin defending the opposing theory of epigenesis. In an accompanying letter, the author, Caspar Friedrich Wolff, admitted that he had read Haller only recently, that he had proposed a theory contrary to Haller's, but that he trusted Haller's humanity and love of truth.[21] It is known that Haller replied appreciatively and offered a few suggestions. Haller also wrote a review for the *Göttingsche Anzeigen*, in which he pulled Wolff's thesis apart and warned that Wolff's theories meant the end of mechanical medicine.[22]

The issue, as both men knew very well, was the central problem in the science of generation, the source of organization attained by an embryo as development proceeds. Was organization, as Haller argued, the manifestation of organs already present, or was it, as Wolff had it, the successive formation of parts from relatively unformed materials? Haller may have hoped the voice of authority would bring young Wolff around. If so, he grossly underestimated his opponent. Doctor Wolff published a sharply worded defense in which he hinted that Haller invoked the initial invisibility of parts rather too often and saw what he wished to see.[23] The ensuing hostilities lasted until Haller's death.

The year 1760 ranked as a fairly successful year for Linnaeus. He published a second edition of the *Fauna svecica* and a plant index for the tenth edition of the *Systema naturae*. His students at Uppsala defended ten of his dissertations. Of these, first place goes to the *Politia naturae*, a somber analysis of the means nature employs to maintain order and balance in her economy.[24] The previous year Linnaeus had submitted two papers to the St. Petersburg academy. One, a short description of *Nitraria*, was now published in the commentaries.[25] The other, submitted in answer to a prize question on plant sexuality, was pronounced "praemio omnino dignus," and a prize of a hundred ducats was awarded Linnaeus at the session of September 16, 1760.[26]

Unknown to Linnaeus, his answer had been closely scrutinized by an adjunct of the academy, Joseph Gottlieb Koelreuter.[27] Stimulated by the prize competition, Koelreuter had set up experiments designed to untangle the problem of plant reproduction. The following year Koelreuter published his preliminary results and, later, three supplements (1763, 1764, and 1766), all of which contradicted Linnaeus's wilder claims and speculative flights.[28] Not content with this, Koelreuter condemned Linnaeus's theory of generation and experimental procedures in the harshest terms.

An attack from this quarter was unexpected. Koelreuter was a student of Johann Georg Gmelin, formerly a Linnaean ally at St. Petersburg and now a professor at Tübingen.[29] Gmelin had been among the very first to recognize the importance of Linnaeus's research on hybrids and at his inauguration in Tübingen had recommended experiments in the new field. Koelreuter's investigations were the result of research that, in the opinion of the experts, "absolutely contradicted the opinion of the Chevalier Linné." Worse still, Koelreuter persisted in his specialty and mastered the field of plant reproduction, a field in which the theories of Linnaeus had by that time only historical interest.

Buffon's experience with the next generation was different—but not better. Almost all university education in natural history was inspired by the idea of rigorous classification. Not only did synchronic essentialist methods work against Buffon's diachronic genealogical method, the format afforded no space for the mention of Buffon's ideas. Only bits and pieces of Buffon's descriptive techniques and his actual dispositions of species found a place in the training of young naturalists. This was the origin of the legend of Buffon's scientific incompetence. Every specialist in some part of the table of organization could cite examples of Buffon's recklessness and caprice.

This reading of Buffon was dominant but not omnipotent in university circles. Independent reactions to Buffon were formed outside Göttingen, Halle, and Leipzig. One positive reaction came from Peter Simon Pallas, who had gone to Leiden to study medicine.[30] In his *Elenchus zoophytorum* (1766), Pallas showed himself surprisingly independent of the older experts, John Ellis and Linnaeus.[31] He shrewdly criticized key essentialist concepts and, even more interesting, had read Buffon. He distinguished between homology ("Struktur und Zeugungsart") and analogy ("superficielle und idealische Verwandtschaft"), for example, along the lines laid down by Buffon between the physical order of bodies and the abstract order of ideas.[32]

Pallas's next books were received as supplements to Buffon, but in fact, Pallas was engaged in the exploration of Buffon's systematic dispositions with an eye to refuting Buffon's wilder generalizations. Some years later, Pallas analyzed and rejected Buffon's concept of degeneration.[33] He contrasted Linnaeus's essentialism with the genealogical method of Buffon. He had, he said, struggled to free himself from the fetters imposed by their authority, and had found that the only sure method was to collect and compare the facts, fix his own ideas on the general causes of variation, and compare these with the opposing doctrines of the two men, who had left their science tied in knots. The three principal causes of degeneration cited by Buffon—climate, nourishment, and accidents multiplied by number—did not change any animal to a degree that would sustain the concept of degeneration. This left one internal cause, reproduction, and here again the effects were less decisive than Buffon allowed. Pallas concluded, "we must renounce this supposed degeneration of the species" (ibid., p. 100).

These younger men, Joseph Gottlieb Koelreuter, Peter Simon Pallas, and Caspar Friedrich Wolff, were members of a generation of scientists born in the 1730s and 1740s.[34] Early in their education, they had oriented themselves toward research careers in natural history and physiology. All three of them found positions in St. Petersburg, and theirs was the first generation to take the rigorous demands and specialized research of academies for granted. All were adept in the rhetoric of the scientific enlightenment. They preferred physical to metaphysical explanations and accepted d'Alembert's distinction between "l'esprit de système" and "l'esprit systèmatique." This did not mean that they had abandoned the closed worlds of rational systems altogether, but the emphasis in their work was on the open horizons of observation and experience. They were specialists, far more so than their better-known elders if not to quite the degree that we now take for granted. And they began their careers around 1760.

The year is important. By accepting 1760 as a turning point in the life sciences, I intend to link these studies with the thesis of a recent collection of papers, *The Quantifying Spirit in the Eighteenth Century*. In his introductory essay, John Heilbron argued that the later eighteenth century saw a rapid increase in the range and intensity of application of exact methods. After 1760 a new emphasis "did indeed come to invest not only the sciences and their applications to fields as distant as artificial languages and forest management. Even in the exact sciences, such as astronomy, optics, and mechanics, we find a newly effective emphasis on precision."[35]

To what degree does this thesis apply to the life sciences? Heilbron valiantly attempted to include all systematized information in his quantitative thesis, but it is clear that a majority of eighteenth-century naturalists and physiologists assumed that mathematics was incompatible with the study of living forms. On the other hand, it is also clear that an important inflection occurred in the sciences of living form around 1760. With important modifications, the idea of increasing precision—not necessarily in the form of strict quantification—applied to the sciences of living form—and on a far greater scale than can be dealt with here comprehensively. Let me offer a couple of striking examples, using research I have already mentioned.

When Joseph Koelreuter set out to demonstrate the limits to crosses between two plant species, he conceived the problem as a combinatory. The parts of each species were susceptible of measure and enumeration; a combination of two species would result in the addition of a sum or the subtraction of a difference for each of these parts. Koelreuter's tools—measurement and counting—were the same as those occasionally used by contemporary systematists; the difference lay in the consistency with which Koelreuter applied these simple tools. His experiments with hybrids were set up on such a scale and carried out in such a way that he could determine the forms and sizes under which hybrid offspring appear, arrange these according to their different generations, and ascertain roughly some of their numerical relations. Koelreuter's results were products of the stubborn persistence with which he aimed at an exact determination of certain limited aspects of plant forms.

Similarly, Caspar Friedrich Wolff recast the generative process as a combinatory involving simple elements and their modes of successive composition. "One truly explains generation," he wrote, "who deduces the parts of the body and the mode of its composition from known principles and laws."[36] To a degree, Wolff forced his observations into the straitjacket of his deductive method. His vessels and vesicles are not to be confused with vessels and cells. They are simple rational elements, building blocks, from which formation was to be reconstructed. Similarly, his agent of formation was not a product of observation. To provide a causal explanation for successive formation, Wolff interpreted the relation between nourishment and structure as a force whose effects separated organic formation from all other modes of composition. On this basis he "explained" the general procedure of production and the organization of parts, the process of vegetation common to plants and animals.

These two unrelated attempts to adapt combinatorial analysis to organic processes are particularly striking instances of the quantifying spirit at work in the life sciences around 1760. This is not to argue, however, that these applications were typical or representative of ordinary procedures in natural history or physiology. Koelreuter and Wolff's scientific contemporaries were masters of impersonal, qualitative description, and throughout the eighteenth century they ignored or resisted the use of mathematical methods and symbols. But even the die-hards found it hard to resist impulses toward greater exactness originating in more prestigious spheres of the scientific constellation. Measurements of size, an old anatomical procedure, became an important element in the fledgling science of comparative anatomy.[37] Naturalists also counted seeds and pollen grains, observed numerical ratios among body parts, noted numbers of progeny, and proposed mathematical nomenclatures. For the most part, though, the increasing emphasis on precision in the life sciences after 1760 had nothing to do with quantification and mathematical methods. We find instead a concerted attempt to make qualitative observations far more exact. Here are a couple of examples.

Albrecht von Haller repeatedly rejected Wolff's version of epigenesis. Wolff argued that the heart, precisely like all other organs, took form gradually. Haller was convinced that the heart had to exist from the beginning of the embryo. Since both men possessed comparable optical resources, Wolff's only recourse was to extend the limits of observation. Since Haller began his observations of incubated eggs at forty-eight hours, Wolff began his observations at twenty-four hours, and he claimed he had seen the origin of the heart and its first feeble contractions.[38]

Haller rejected the observations. If, as Haller argued, the invisibility of the heart did not demonstrate its nonexistence, Wolff had in a sense to prove nonexistence. In his treatise on the formation of the intestines, Wolff did so in two different ways. He showed that there are modes of formation not to be reduced to any scheme of preformation: the closure of an organ originally open and the fusion of two sheets in a tube. Wolff also showed that formation operated by substituting part *B* for part *A*. *B* could not have been present from the beginning, since *A* occupied its place.[39] The exactness of Wolff's observations did not convince Haller, of course, but a century later, Karl Ernst von Baer pronounced Wolff's work "the greatest masterpiece we know in the field of observational science."[40]

A second example of increasing precision in observation comes from the very different field of taxonomy. By 1760, it was obvious that the natural method for plants would rest upon the function of fructification. Since the time of Caesalpino, naturalists had equated fructification with the essence of the plant and had extracted differentia from the complex organs devoted to this function without investigating very thoroughly any part at all difficult to observe. Travelers who reported on distant floras neglected to gather fruits or did not know what plants those that they had collected belonged to. European collections were full of unknown fruits.

When Joseph Gaertner (1732–91) and Antoine-Laurent de Jussieu (1748–1836) conceived their methods independently, both foresaw that the key to the natural orders would be a comparative anatomy of fruit and seed. Gaertner in particular pushed this microscopic analysis to a degree of exactness and thoroughness unsurpassed during the rest of the century.[41] "We may state," pronounced Georges Cuvier, "that in all the eighteenth century, there was not a single work that equalled Gaertner's for the profundity of investigation or for the great number of species for which he had observed the fruits and seeds."[42] Jussieu received Gaertner's first volume just as he was printing the *Genera plantarum* in Paris in 1789. He added a note in which he related his genera to those Gaertner had established under another name. In a series of memoires published long after the French Revolution, Jussieu reworked his natural orders, controlling the characters and compositions of each by means of Gaertner's meticulous analyses and figures.

These examples of increasingly precise observation in the life sciences after 1760 are not just circumstantial and arbitrary. The triumph of epigenetic theory in the science of generation and the discovery of a natural method in botany were two of the great achievements in physiology and natural history between 1760 and 1790, and they had consequences for many other disciplines in the life sciences. At the very center of these achievements was the newly effective emphasis on exactitude, thoroughness, and precision.

The rise and, more importantly, promotion of exact methods in the sciences after 1760 indicate broad social support. Heilbron has mentioned wider social movements possibly associated with this change: rationalizing bureaucracies, incipient industrialization, and the rising tide of democracy. A significant factor in the story I propose to tell here is the rapidly increasing number of posts for salaried academicians. After 1750,

> the rate of founding of academies doubled over what it had been in the previous half century. This enlargement continued until the French Revolution and affected the quality as well as the quantity of academicians. The insistence upon expertise in a science as a condition for admission, although not always imposed, did create an environment more favorable than earlier more lenient times to the advancement of quantifying spirits.[43]

Research at universities and medical collegia was affected by the new insistence upon precision. Professors were often academicians or had colleagues who were. With research went instruction, and in short order the new methods reached fields devoted to the applications of science, government bureaus, and the swarm of private societies promoting science of all kinds. One obvious result was an enormous increase in the production of specialized investigations in every known discipline and in fields never before imagined.

At the founding of the Swedish Academy of Sciences in 1739, Linnaeus had praised academies. United forces, he said, act more strongly; when something great is to be accomplished in science, it can be done only with the help of many. The study of nature was man's purpose, and the study progressed where the forces of many prised the stone.[44]

But there was another side to the matter. Linnaeus was also to experience the truth of another of his sayings. "Science, like *Cynosurus coeruleus*, withers where it began but spreads all around."[45] Linnaeus had offered science the one true doctrine, and he lived to see his dogma questioned.

In his later years an anti-Linnaean faction grew up in the academy he had helped found. Linnaeus, said Charles de Geer, was blinded by his own opinions; Peter Jonas Bergius and Torbern Bergman openly disparaged the master's science; and Pehr Wargentin wrote Haller that the members of the academy valued Linnaeus, "but hardly anyone loves him, not even here."[46] Linnaeus claimed to be indifferent. His consolation was that the greatest botanists of his generation had accepted his teaching.

Albrecht von Haller experienced the same indignity. Even with the full weight of his authority, he failed to silence Caspar Friedrich Wolff. In fact, Haller and Charles Bonnet did more to broadcast Wolff's ideas than Wolff himself. They repeatedly refuted Wolff's brilliant observations and continued to insist that embryonic organs were present but invisible. But Wolff had the last word: six years after Haller's death, Wolff wrote Leonhard Euler that Haller had boasted he would silence Wolff. Then, wrote Wolff, "I wrote the dissertations on the formation of the intestines. Haller said they were important but never responded to them. It seems, then, that M. de Haller was quieted down himself."[47]

Generation was only one of Haller's interests, but in other areas in which he involved himself, such as the origin of the heartbeat, animal heat, and electrical impulses in nerve processes (which he dismissed), younger men also questioned his pronouncements. At Göttingen, Johann Friedrich Blumenbach rejected Hallerian preformation and even dared to modify the research on irritability and sensibility. "Haller, the great arbitrator in the doctrine of irritability, has ascribed it improperly, I think, to some parts possessed of contractility, but in which I have never been able to detect genuine irritability."[48]

Buffon's natural history, although a great popular success, also did not fare well with scientific contemporaries or their immediate successors. Detractors contrasted Buffon's sweeping views with the patient observations of Réaumur or the younger Charles Bonnet. Malesherbes criticized Buffon's ideas on classification, Condillac his system of animal faculties. D'Alembert called him "le grand phrasier." "Buffon," said Marmontel, "had the chagrin to see that mathematicians, chemists, and astronomers accorded him only a very inferior rank among themselves, and that naturalists were very little disposed to place him at their head."[49] In the Reich, where Buffon's natural history continued to be dismissed, he did attract one influential colleague whose support proved decisive. Immanuel Kant understood, as naturalists did not, Buffon's distinction between abstract and physical systems, and in 1790, when Kant wrote the period's most influential analysis of the life sciences, he brought the ideas of Buffon to the attention of scientists in a way that could no longer be ignored.

In the second chapter of Buffon's treatise on generation, there is a section on the framing of scientific questions that deserves to be better known. One must distinguish, says Buffon, questions of why, how, and what. When a savant investigates reproduction, say, he does not ask why organisms reproduce: the question is insoluble. Nor will he ask how organisms reproduce: that question results in a list, facts without causes. But if the savant asks what means nature employs in reproduction, he has a question that can be investigated. The question is neither insoluble, that is, dependent on first causes, nor is it a pure question of fact. The question relates reproduction to analogous natural processes from which hypotheses can be drawn. "The question itself allows us to make hypotheses and to choose that which seems to have the greatest analogy with the other phenomena of nature."[50] For all its brevity, his discussion states the program of the scientific enlightenment in the life sciences. Living nature was to be explained in her own terms and as a unified and self-contained system.

When we turn from the high and wide world of programmatic statements to their realization, however, the picture changes. The inquiries of Buffon and his generation claimed to explain particular effects and processes, but they found themselves obliged to reach out time and again toward final causes and immutable plans. The equivocal and contradictory results were an important part of their legacy.

Their successors, men who began their careers around 1760, were far more consistent in framing their questions. Rather than attempt frontal assaults on the system of nature, the sources of structure, or the causes of diversity in the manner of the triumvirs, the men of 1760 preferred to serialize their problems so as to seek a particular solution for each of them. And when they approached a particular problem, they avoided fruitless speculation and contented themselves with attainable goals. Pallas spoke for his generation when he stated his preference for limited problems and accurate, well-coordinated information.[51] In their insistence upon secondary causes and intelligible principles, the generation of 1760 backed the scientific program of the enlightenment with a tenacity and rigor not found in earlier and more lenient times.

An established topic in the rhetoric of the earlier enlightenment was the harmony of reason, revelation, and experience. "Truth," wrote Haller, "is supported on all sides; everything conspires to support her edifice."[52] On this basis Haller professed indifference to the success or failure of his opponents' theories. "New light will always bring us closer to the truth and to God by the means of truth" (ibid., p. 67). But the record shows that Haller was anything but indifferent to the success of his opponents and that he was painfully aware of the many instances where reason, revelation, and experience did not pull together. This was true of Linnaeus and

Buffon, as well. Early in their careers, both men had run afoul of theological faculties and had responded with public disclaimers and private contempt. Throughout their careers they repeatedly ran up against discrepancies among kinds of truth, and as we shall see, their works offer many examples of compromise among the conflicting claims of received ideas, rational abstractions, and observed facts.

Their successors were far more likely to admit the discrepancies openly. It was not that the younger men had succumbed to materialism and atheism; they were theists and deists of various kinds in their officially Christian societies, but they subordinated belief to their scientific vocation. And when they presented evidence that contradicted revealed truth, they did not expect, nor did they receive, clerical scrutiny. In 1778, Eberhard Zimmermann proposed a version of organic distribution that replaced the Garden of Eden with multiple centers of creation. He candidly admitted that the idea could not be reconciled with the "tale" found in Genesis. But this, he said, was not to be expected in an enlightened age.[53] This is only one instance of the relative autonomy achieved by the sciences of living form after 1760. There are other, even more important instances, and to these my story now turns.

II

The System of Nature

The *Politia naturae* of 1760 offers an example of the paradoxes Linnaeus expounded in his academic dissertations. Let us suppose, Linnaeus began, that a man as naked as Adam at the Creation was suddenly set down on our terraqueous globe. If he attended carefully to the spectacle before him, he would soon discover that earth's innumerable creatures crowd upon and conflict with one another and survive only by preying upon one another. Observing this war of all against all, he would feel threatened by violence from all sides. But after he had dwelt here for some time, he would begin to detect here and there in the apparent disorder the links of a chain of order so artfully composed that "we can scarcely discover among the divine works a beginning and an end."[1] Linnaeus used the rest of the dissertation to demonstrate the necessary connection between appearance and reality. The merciless struggle for existence was a prerequisite for the functioning of creation.

My concern here is not the paradox, nor the dissertation, but the assumption underlying both, the affirmation of a natural order, a system of regular, constant, and universal relations among created beings. This was a belief that Linnaeus shared with his audience at Uppsala and with almost all of his European contemporaries. The idea could appear in very different contexts and serve very different purposes. Apologists used natural order to justify the ways of God to man. Philosophers used it as a basis for their deductive methods. Naturalists used it to arrange observational data. Whether men busied themselves demonstrating the continuity of creation by comparing natural forms or whether they described the forces that maintained the equilibrium of the divine economy, their assumptions about the regularity and permanence of relations among created forms constituted the value of their explanations.

Unlike the views that prevailed a century later, natural order was not regarded by most naturalists of the eighteenth century as essentially the product of physical and temporal processes. Natural order was rather an a priori principle that regulated those processes and constituted their reason for being. The greatest expert on this difficult question in all Europe was Christian von Wolff at Halle. In his *Philosophia prima sive Ontologia*, Wolff had demonstrated to the satisfaction of many of his academic readers that there was an order of things, ultimately founded on the nature of God. In this order, everything had a reason for being, and its existence could be fully understood only when its sufficient reason was disclosed. This sufficient reason was to be sought not in things themselves but in the system to which they belonged. The order of things separated truth from fantasy, and so truth, too, was rooted in the principle of sufficient reason. The inner order of truth had to be distinguished from the outer order of space and time, both of which presupposed the inner order. Space provided the order for the outward existence of things, time for their succession. The order of things was the basis for all occurrence, all knowledge, and all perfection.[2] Order, conceived as prior to and inherent in the system of relations that constituted it, seemed to preclude the possibility that natural order was the product of random processes or that the system had formed itself over time.

The conception of natural order as one vast system of interrelated forms was particularly apt to foster teleological interpretations. The very idea of a system defined nature as purposeful even when nothing was said about the purposiveness of the forms arranged in this way. Nature, regarded as unfolding according to the law of continuity, was regarded as proceeding according to a rule. But of course almost everyone was willing to pursue this line of thought much further. Those things that in themselves lacked a purposeful form could be drawn into the course of nature through a network of reciprocal uses. As for living beings, each kind was treated as a superior artifact, designed to function independently while contributing to the system as a whole. A design always preceded the use of materials; an established course was independent of its realization. Nature everywhere was regarded as subject to a plan, as more or less consciously oriented toward a goal.

The teleological element in the idea of natural order was a mainstay in the apologetics of natural and physicotheology. This is an enormous literature of shifting quality and interest, whose vogue long outlasted the eighteenth century. Its general history has yet to be written—and is also outside the scope of this study. Here I shall content myself with sampling opinion on a single idea that had a profound impact on the naturalists who are the object of my study.

"Nothing takes place suddenly," wrote Leibniz in the *Nouveaux essais*, "and it is one of my great and best-confirmed maxims, that nature makes no leaps. I called this the law of continuity."[3] Leibniz added that there was much work for this law in the natural sciences. And in fact continuity did prove to have many applications, particularly in mathematics, where it led Leibniz to the differential calculus. In the life sciences, Leibniz himself assimilated the law of continuity to the old idea of synexia, a series linking natural forms by multiple affinities in a continuous sequence.

By 1750, scientific publicists were using the law of continuity to create a characteristic *Naturphilosophie*. August Thienemann has shown, for example, how Charles Bonnet, in his *Contemplation de la nature*, used the law of continuity to justify his *gradation naturelle* in the three kingdoms of nature.[4] Natural order was the product of eternal reason, which had chosen to exemplify the kinds of things so exhaustively that no genuine potentiality of being went unfulfilled. The order of these things was known in outline. Creation proceeded little by little from things lifeless to animal life—and beyond—in such a way that it was impossible to determine exact lines of demarcation. The *Contemplation de la nature* was an international success. Following the appearance of Professor Titius's German translation in 1766, a number of German treatises reflected Bonnet's quasi-theological approach to continuity or borrowed his rationale to justify the construction of a single series of natural forms.[5]

Bonnet's construction did not pass unchallenged. If, argued Jean-Baptiste-René Robinet, we admit the law of continuity, "we ought likewise to admit all that follows from it, while it is inexcusable to abandon so general a principle without a sufficient reason."[6] But Bonnet had done just that when he introduced the attribute of reason as applicable at one end of the scale but absent at the very next level. Many of the differences used by Bonnet turned out to be differences in kind, not degree. The law of continuity led, then, to the recognition of radical discontinuities for which there were no links. There was no middle sort of being, for example, between organic and inorganic. Among many other consequences, Robinet eventually projected the development of preexisting germs along a temporal axis in such a way that the simplest beings appeared first, and the history of life seemed to be the history of increasing complexity.[7]

Another school of thought on continuity stressed its physical nature. Space and time in themselves were abstract ideas; space and time in relation to the natural world inhered in the ordered coexistence and succession of created beings. Continuity was precisely the series of beings, either coexisting or successive, chained to one another. There was also a kind of truth in this world, because nothing was made without a sufficient reason and because nothing contradictory found a place here. The mechanism of the world was the source of its truth, a relative, physical truth, but one that distinguished this from a purely imaginary world. Buffon argued that continuity rested upon "the constant succession and uninterrupted renewal of the individuals who constituted it," and this series produced a physical truth. Physical truth was only probable, but the probability was "so great that it is equivalent to certainty."[8]

Ideas as abstract as continuity would seem to have no place in the workaday world of science. And it is the case that the naturalists who continued and reformed orthodox tradition were preoccupied with questions of a much more practical nature. But their daily round—the identification of mineral, plant, and animal specimens and their arrangement in a table of organization—actually supported abstract ideas like continuity. The rules, forms, and models of classification transformed natural objects into ideal beings, arranged on the basis of purely formal affinities and analogies. To be sure, the various methods of classification resulted in clear-cut groups of forms that contradicted the shadings and overlappings so confidently affirmed by the philosophers. But these discrete groups only reflected the inevitable imperfections of classification and were considered provisional. If nature made no leaps, then any hierarchical method interrupted natural order. Around 1750, many naturalists doubted that any progress toward a natural method would be made by dividing and ranking natural forms.

During the closing years of the eighteenth century, however, learned opinion began to reverse itself on this point. First botanists, then zoologists, began to perceive that a hierarchy of groups within groups actually represented formal relations found in nature. Naturalists, who had long distributed classes in orders, orders in genera, and genera in species, gradually became convinced that the procedure was potentially an accurate representation of natural order. The perception was fundamental not only to systematics but also to comparative anatomy, embryology, paleontology, and even biogeography.

In terms of theory, this was indeed a reversal. Since the late Renaissance, naturalists had used the general divisions of forms—the classes and orders—to impose a framework upon their field of study, dividing natural forms into more manageable, approximately equal sections. And then gradually, during the later eighteenth century, increasing numbers of naturalists realized that it would be possible to establish the higher classes of plants and animals upon a natural, physical basis.

Nearly seventy years ago, Henri Daudin demonstrated the importance of this reversal in detail, particularly with respect to the work of naturalists at the Muséum d'Histoire Naturelle.[9] Daudin's argument has left few traces in subsequent research, perhaps because the subject seems unnecessarily abstruse and remote from the interests of working naturalists. But this neglect has meant that the significance of the natural method is no better understood now than at the time it was worked out. When the shift in informed opinion took place, between 1789 and 1795, naturalists began to arrive at the insight that the pattern of groups within groups was, at least potentially, an accurate representation of natural relations. But they offered very little justification for their discovery. Although they understood that a reform in the method of classification had taken place, for some of them this meant only that they no longer had to choose between their formal discipline and the natural method; the pattern of classification had become the pattern of natural order. A much smaller number understood that the natural reality of orders and classes of plants and animals implied a physical basis, a unity of plan with determinate physical effects, and that this unity itself required a physical explanation.

I have taken over Daudin's argument but have established it on a different basis, using the evidence provided by natural history in northern and central Europe. Following Daudin, however, I have concentrated upon research concerning the upper reaches of the systematic hierarchy, the classes and orders of plants and animals. Then, as now, the methods for determining these primary divisions differed from those used to characterize species and genera. Within these limits, I shall argue that the natural method reached something like a definitive form between 1789 and 1795 and that this purely systematic study involved other, very different kinds of knowledge, which eventually transformed European conceptions of natural order.

The Linnaean reform of natural history was most explicit in the science of botany, where, as Frans Stafleu has shown, the prescriptions of *Fundamenta botanica* and *Philosophia botanica* had a decisive influence in all of Europe.[10] Between 1735 and 1778 an army of professors, academicians, and amateurs in Austria, Germany, the Netherlands, and Switzerland evaluated, extended, limited, and refuted Linnaeus's artificial sexual system of classification. As commentary accumulated year after year, Linnaean descriptive models established themselves more and more firmly in botanical practice. Linnaeus's nomenclature was so useful, his method so simple, that botany became a kind of game. Princes rearranged their gardens, professors reworked the national floras, and amateurs everywhere took up botany.

The triumph of the Linnaean method is a success story almost unparalleled in eighteenth-century science. Until quite recently, though, there has been little interest in the factors that entered into that success. Historians of science have often dismissed the work of Linnaeus as old-fashioned and unscientific, so it is difficult to explain just why so many contemporaries received Linnaeus's qualitative method so enthusiastically.

The success of the Linnaean method was a consequence of the fact that Linnaeus was an ingenious and authoritative organizer who attacked the problems of classification on a broad front.[11] Many resisted the standardization of their discipline, but even the recalcitrant recognized that Linnaean systematics enabled naturalists to work together as a community: "As a system of citations and technical language, it [the Linnaean system] performed the critical function of joining naturalists together in a working community."[12] As for scientific amateurs, Linnaeus's presumption of the validity of the world as man actually experienced it was an element in the appeal. The method constituted an alternative science, one around which antiphilosophes and antireductionists could rally.

There is no evidence, though, that naturalists ever considered the Linnaean system the culmination of science. "As little as I believe," wrote J. C. P. Erxleben,

> that the essential element in natural history consists in the marks of classes, orders, genera, and species of natural bodies (for I regard these simply as a guide to prevent us from going astray in the extent of the science), so, too, I cannot regard Linnaeus's system of nature and comparable books as true compendia in natural history; they are only registers of known natural bodies.[13]

Since, however, the application of the Linnaean method to problems of identification proved, above all other methods, easy and intelligible, many naturalists found nothing more useful than to perfect and extend parts of the Linnaean method originally defective or wanting. Early German discussion focused upon the defense of plant sexuality and upon rationalizing a somewhat artificial classification.[14] Others tried to extend the theory of plant sexuality to the cryptogams.[15] The gradual improvement and extension of the system as a whole went on long after the death of Linnaeus and is best seen in the many editions of the *Species*, *Genera*, and *Mantissa plantarum*.[16]

The study of natural affinities in plants was considered an entirely separate affair. "I do not," Erxleben wrote in 1773, "regard the Linnaean system as much more than a tool of identification." But, he continued, "doubtless we can imagine a method based on an arrangement according to the greatest likenesses in natural bodies where those bodies are placed next to one another that agree in most of their qualities."[17] This natural method would classify plant forms in such a way that their actual order would be condensed and reproduced in the method. Comprehending all plant parts, properties, faculties, qualities, and virtues, the method would admit no exceptions and would be independent of the practical interest or value imputed by human will. But like the philosophers' stone, the natural method would also offer a key to relations with which to unlock the uses and remedies of plants.[18] Almost all European botanists regarded the natural method as the ultimate goal of their science.

Essays in the natural method from the mideighteenth century are dominated by the discovery of simple basic elements and their progressive combination. Parts of flower, fruit, and seed appear over and over again, now characterized by one dimension, now by another. The need to stabilize class and ordinal concepts transformed the ideal of essential definition into mere combinations of abstract characters. When we examine the principles of these combinations, however, we find the same ideas that dominated Linnaeus's thought on the natural method. Nature forbade the exclusive use of single characters, which was true of artificial methods. The natural method was supposed to account for all the affinities among natural groups. On the other hand, botanists knew that not all likenesses among plants were equally important; some could be neglected in comparison with others. The successors of Linnaeus did not opt definitively for a complete inventory or for a hierarchical selection of characters. They used these seemingly antagonistic ideas as complementary principles for dealing with the mass of plant forms.

As examples of this way of thought, let us turn to three methods published between 1748 and 1768, Johann Gottlieb Gleditsch's "Système des plantes," Albrecht von Haller's arrangement of Swiss plants, and Lorenz Heister's *Systema plantarum generale*.

After 1730, Lorenz Heister pursued an arrangement of plants in the gardens of Helmstedt that had less to do with genera and species than with an accurate representation of affinities in the larger groups of plants. Heister published this arrangement in 1748, ten years after the appearance of Linnaeus's enigmatic *Fragmenta methodi naturalis*. Heister began by dividing trees and shrubs, with the result that he separated many closely related forms. After this first cut, however, he established three legions, two based on the number of seed leaves (mono- and polycotyledons) and a third upon the obscurity of fructification (cryptocarpa). Heister distributed these legions into classes, orders, genera, and species on the basis of flower, fruit, and seed.[19] Heister's arrangement did not attract much attention in the Reich, but according to de Candolle, his pamphlet guided Bernard de Jussieu when, in 1758, he disposed the garden of the Trianon into sixty-five natural orders.[20]

Albrecht von Haller's arrangement of Swiss plants in the *Historia stirpium indigenarum Helvetiae*, 1768, established thirteen classes of flowering plants, six of cryptogams. Haller divided flowering plants according to the presence or absence of a corolla. He divided apetalous plants into those with colored, green, or glumaceous perianths, and those whose perianths were replaced by a spathe. Petaled plants were monopetalous with either didynamic stamens or petals borne on the seed, or they were polypetalous, and divided, according to the relations of stamens to corolla, into poly-, diplo-, iso-, or meieostemones; two other classes corresponded to tetradynamia and papilionacea. Haller divided these classes into many sections using a complicated set of characters drawn from flower, fruit, and seed. In his preface, he called these groups families and said he had tried to make them conform to nature without pretending to establish a natural system, which he thought not possible given the limited number of plants considered.[21]

Gleditsch published a very different approach to the natural method in the *Histoire* of the Berlin Academy of Sciences for 1749, his "Système des plantes fondé sur la situation et liaison des étamines." Following Linnaeus, Gleditsch located the essence of the plant in the fructification and the essence of the fructification in the stamens and pistils. Each of four floral parts—calyx, corolla, stamens, and pistils—he considered first with respect to its species then under the four Linnaean dimensions of number, figure, proportion, and situation. Gleditsch worked out each of these as a possible system to discover which of them produced the highest number of natural groups. He arbitrarily limited the plant parts and the points of view from which they were considered and, within these limits, accorded an equal degree of interest to each of the parts. Of course, he found none of his systems satisfactory.

> Although the natural character of plants preserves its force in each of these twenty methods, and although traces of the natural method are found in each of them, nonetheless there is not one that coincides perfectly with the natural method, but all are visibly artificial. This does not astonish me; for this is the common end not only of methods invented by botanists but also of all those that we employ in other parts of physics; no one that I know has found the true key to the natural method in the three kingdoms of nature.[22]

The belief implicit in Gleditsch's statement, and common in research on the natural method during the mideighteenth century, was that an accurate representation of natural affinities would not finally prove to be a classification. Nature, who made no leaps, had not established the divisions imagined by naturalists; her order was that of a continuous series of forms without cleavages. It followed that human classifications were not hers. The attitude was convenient but dangerous in its consequences.

Physicotheologians, without regard for the needs of science, were free to draw up tables of affinities on metaphysical principles, while naturalists, freed from the need to respect continuity, substituted an apparatus of definitions for the study of natural objects. The study of botany was threatened by a divorce between the procedures of identification and the investigation of natural order.

Works by two naturalists of the next generation, Joseph Gaertner's *De Fructibus et seminibus plantarum* (or carpology) and Antoine-Laurent de Jussieu's *Genera plantarum*, preserved botany from this danger by showing how the forms of classification could be conciliated with natural order.[23] Both men were convinced that the achievement of a natural method would transform botany from a science of memory and nomenclature into a new science "with its combinations and affinities as in chemistry and its problems as in geometry."[24]

When Joseph Gaertner embarked upon the comparative anatomy of fruits and seeds at the beginning of 1769, he was an academician in St. Petersburg charged with the direction of the gardens and the cabinet of natural history.[25] His duties left little time for research. Eventually, he resigned his position and returned to his native Calw, in Württemberg. Although Calw offered solitude and the opportunity for contacts with colleagues in London and the Netherlands, it was not rich in resources. The fruits Gaertner amassed were only a small part of those he hoped to examine. When he heard that Joseph Banks had returned from his voyage around the world with immense collections in natural history, Gaertner set out for London. Banks proved unbelievably generous. He allowed Gaertner not only to see and draw but also to dissect all the fruits he had collected; Banks also made a present to Gaertner of his duplicates and used his influence to procure from others what he did not possess.

On his return to Calw, Gaertner stopped in Amsterdam, where Carl Peter Thunberg had just returned from his voyage to the Cape of Good Hope and Japan. Thunberg presented Gaertner with many exotic fruits and later sent many more on to Calw. In Leiden Gaertner again found many objects of interest. In the end, he retreated to Calw, loaded with treasure and ill from overwork.

The carpology underwent more or less continuous revision over a period of twenty years. Descriptions were never sufficiently comparative or exact. The preface was not exempt from speculation. While the engravings were being completed, Gaertner recast the entire text of the first volume. This volume, dedicated to Joseph Banks, was published in Stuttgart at Gaertner's own expense in March 1788. (The second volume was published at Tübingen in 1791.) Thus twenty years of labor resulted in two quarto volumes containing scrupulous descriptions of the fruits and seeds of more than two thousand genera, arranged according to a new carpological method, and preceded by an account of carpological morphology.

Gaertner's goal was the discovery of nature's own system, and he criticized the traditional comparison of natural order with a chain whose links represented plant kinds. Since the key to the natural method could not be found in any one part, it followed that the order of nature should not be represented as a single chain. "The organization of vegetable material," he wrote, "undergoes such diverse modifications, and these modifications in turn yield such a great number of combinations, that the affinities of plants can by no means flow from one source; neither, likewise, can the genera succeed one another in a direct and continuous series."[26] He recalled Linnaeus's earlier comparison of natural order to a genealogical-geographical map. Genera, Gaertner continued, "ought to be placed next to one another like tracts of land and provinces on a geographical map." This map would reveal that acotyledons often invade the realm of polycotyledons, monocotyledons the realm of dicotyledons. The map would also reveal "very tight partial chains of affinities among plants," and these chains of affinities "would in turn yield natural families" (ibid.).

Gaertner also rejected the traditional precept that a natural method could be founded on the consideration of all parts. On the contrary, a natural method could be based only on essential or constant parts. And characters taken from fruit and seed were preferable to those taken from the flower in determining the natural limits and connections of genera. Furthermore, initiates would proceed to the innermost recesses of the seed, where one found those characteristic signs that one sought in vain in the exterior. These signs "surpass others in their constancy and prominence, and when they are neglected or little esteemed, not only are the limits of genera rendered obscure or precarious, but false species may even be confused with true ones" (ibid., p. clxxvi).

As a systematist, Gaertner settled for an artificial, purely carpological method, which, he argued, would approach a natural method more closely than any other system based on the consideration of a single part. At the level of the species, the affinities from which genera were formed required a close scrutiny of the fruit. Again, in determining natural affinities among genera or families, the structure of the fruit always had to be considered. The fruit not only exhibited family characteristics but determined the closer or more remote nexus among orders. Thus, Gaertner concluded, "a universal chain of plants, or a natural method, cannot be built on this [the structure of the fruit], yet it allows us to discover new and partial connections, and to better comprehend those already known" (ibid.).

It was obvious that signs taken from the fruit were not equal in impor-

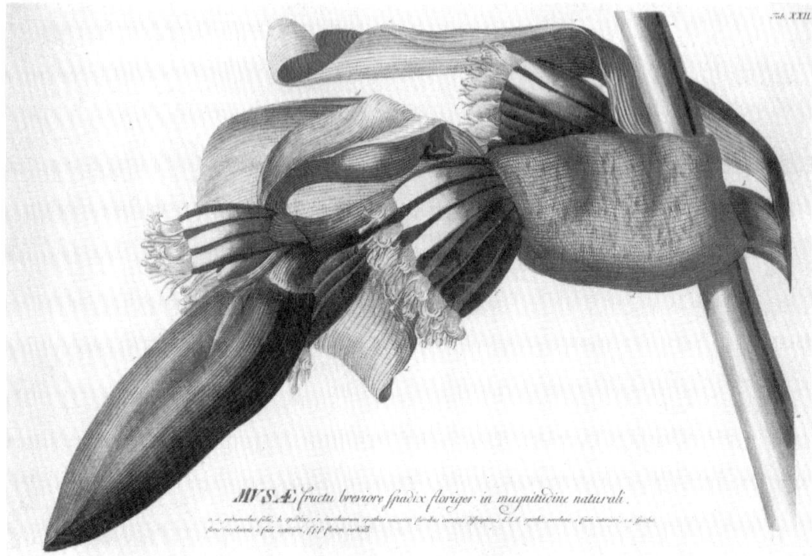

An example of the complexity of fructification: the banana inflorescence by Georg Dionysius Ehret, 1750, from Christophe Jakob Trew, Plantae selectae *(Nuremburg: Felssecker)*

tance. Greater weight had to be given essential or constant signs than to accidental or minor ones. Gaertner established two orders of characters: those deserving little or no attention in constituting genera and those meriting great or extraordinary attention.[27] When parts among characters deserving great attention differed, he wrote, "genera ought to differ also, even when all other remaining parts are similar."[28]

In his classification, Gaertner first distinguished the cotyledons, noting that the division was not always neat. The considerations he then employed were, in order, the position of the fruit, the direction of the radicle, whether the fruit was simple or composite, whether the seeds had albumen or not, whether the embryo was straight or curved, and the consistency of the pericarp and its dehiscence. Throughout his divisions, Gaertner preserved an old-fashioned symmetry. Each of his divisions presented the same cleavages, regularly repeated. Groups at any one level, whether class, order, or genus, were supposed to be equal in weight and value because they were characterized by similar variations.

Gaertner's exacting and thorough work found an appreciative audience at the Paris academy. Lamarck made use of Gaertner's plates in his own *Illustrationes generum*. The academy cited Gaertner's work among those most useful to science in the last few years.[29] The most enduring response, however, came from Antoine-Laurent de Jussieu. The year after Gaertner's first volume appeared in Stuttgart, Jussieu published his own *Genera plantarum* under the privilege of the Académie Royale des Sciences. Jussieu's classic developed independently and confirmed many of Gaertner's principles.

To attain a truly natural order, Jussieu, like Gaertner, argued that the observer must proceed from what was simple and known to what was composite and unknown. Species "alike in the highest degree" and "conforming to one another over a continuous series of generations" were simple by universal consent and by "the most clearly perceived law of nature." When defined by their proper signs, species determined connections with similar species. A genus was thus a congeries of species connected at an imperceptible nexus but without strict limits.[30] After related species were assembled in their natural genera, related genera were in turn assembled in orders, and these orders in classes. "For Nature, not diverging from her established law, always assembles various small groups of plants into larger ones, which by repeated connection are themselves enlarged, until at last she brings all together in one embrace" (ibid., p. xlii).

Jussieu alluded too to Linnaeus's analogy between natural order and a geographical map. A natural method connected plants of all kinds by an invisible bond, somewhat like a chain—but more like a geographical map—"in which species, like districts and cantons, are distributed into territories and provinces or kingdoms" (ibid., p. xxxvii). The image showed that individual species, as well as groupings of genera and classes, were uniformly coterminous with those surrounding them.

Jussieu acknowledged that the materials of natural order were difficult to assemble and that absolute knowledge of all existing plants was not to be hoped for. The principles that served as the basis for a representation of natural order, on the other hand, were easily recognized. Almost all botanists agreed, for example, upon some natural genera. These, when analyzed, revealed "the true mode of generic construction" and could be used to work out other genera less distinctly indicated (ibid., p. xxxviii).

Similarly, there were some incontestably natural orders. Jussieu believed that these orders offered a key to the natural method. By comparing the characters of a known order with the genera that composed it, a botanist could distinguish ordinal from generic characters. By comparing orders, he could establish the relative importance or constancy of ordinal characters and, in this way, arrive at a value for each character. This value could then be used to determine less clearly indicated groups. The comparative analysis of these groups would result in new rules, whose application would result in new orders. This, he concluded, "is the only rule governing the creation of orders, the only means of discovering affinities" (ibid., p. xlii).

Jussieu had even less use than Gaertner for a uniform consideration of all parts as a basis for the natural method. Throughout his analysis, Jussieu used a principle that came to be called the subordination of characters. One of its bases was functional importance. In plants, there were two functions, nutrition and reproduction, and hence two systems of organs, vegetation and fructification. Within each system, the primary organs serving these processes were "rightly set before all others" (ibid., p. xix; see also pp. xl–lii). But since the number, form, and proportion of these organs in themselves contributed little to the function, signs taken from these modes were "less to be valued on account of their instability" (p. xix). Consequently, the second basis for the principle of subordination was constancy, and characters were evaluated on the basis of their relative instability.[31]

With this rationale, Jussieu established three classes of characters. The first were essential, that is, drawn from essential organs, primary and constant. Examples of these were insertions of the stamens, the mutual arrangement of stamen and pistil, and the number of lobes in the corculum. The second were generally present and varied only as an exception; these were drawn from organs that were not essential and even occasionally lacking. Some examples of these secondary characters were the presence or absence of a perisperm, a calyx, or a nonstaminiferous corolla, the structure of the corolla, and the mutual location of the calyx and pistil. Other, tertiary characters were now constant, now variable, in the orders and were drawn from essential or nonessential organs. A calyx, for example, might be mono- or polyphyllous, an embryo single or multiple, and the number, proportion, and connection of stamens might vary. Genera of a single family necessarily conformed in characters of the first class, generally conformed in characters of the second class, and often conformed in characters of the third class.

The importance of the principle of subordination rested upon the combination of characters in each order of plants. A character of the first class entailed a certain number of other characters. To state that this character was present was to establish the coexistence or absence of others. The presence or absence of cotyledons, for example, manifested itself in all parts of the plant that presented profound differences. To state that a plant was mono- or polycotyledonous was to state many other facts about the plant: the general agreement of elementary organs, the mode of germination and ramification, the structure and nervation of the leaves, the symmetry of the flowers, and so on. Similarly, from a character of the second class, a botanist could deduce many other characters of a superior, equal, or inferior group. To state that a corolla was monopetalous was to state that a plant was dicotyledonous and that most of the stamens were inserted on the corolla and were equal or inferior in number to the corolla's divisions. Knowledge of these predictable relations among parts was the key to the natural method. The more constant and essential characters, Jussieu wrote, "work together or are connected to one another in such a way that the presence of one always announces that of the other" (ibid.).

One of the physical bases for Jussieu's orders was the comparative anatomy of seeds. In a memoire published in 1775 Jussieu noted "that all the seeds of a plant and all those of a genus have an embryo situated in the same manner; this conformity is encountered quite generally in all plants of families recognized as very natural."[32] This structure, alike in many families, led Jussieu to suspect that it would be equally so in all, and he spent months dissecting seeds taken from collections in Paris. He was one of the few men in Europe who could appreciate the importance of Gaertner's first volume, received just as he was printing the *Genera plantarum*. He added a note in which he related his genera to those of Gaertner. Later, he admitted that Gaertner's observations, added to the characters of the *Genera plantarum*, "would have resulted in many changes in the relations of genera and the dispositions of families" (ibid., p. 220). In an extended series of memoires published after the French Revolution, Jussieu reworked his natural orders, using Gaertner's analyses and figures.

The agreement between the independent observations of Gaertner and Jussieu and the conformity of principles by which they judged the value of plant characters seemed to prove that they had followed the path indicated by nature. Paul Dietrich Giseke's belated publication of Linnaeus's lectures on the natural method in 1792 signaled a renewed interest in the reconciliation of the formal procedures of classification with natural groups.[33] But for all the points of agreement among the methods of Linnaeus, Gaertner, and Jussieu, it was clear that Linnaeus's information was out of date. The carpology and the *Genera plantarum* had produced a reform in botany. "The one and the other," Georges Cuvier wrote, "have led to all of the botanical works that have been published since."[34]

The significance of that reform is difficult to assess. Gaertner and Jussieu were certainly not the first to demonstrate the physical existence of some plant genera and orders. Some of these groups had been recognized since the sixteenth century. Nor does the importance of the reform lie in a radical change in classificational procedure. The hierarchy of concepts and definitions established by Linnaeus remained intact. The principle of subordination implicit in Gaertner's work and stated openly in Jussieu's

Genera plantarum was only an adaptation of procedures used by European naturalists since the time of Caesalpino. Logical division, as it was understood by these men, adopted a hierarchy of classes in order to reduce the number of considerations needed to define a group of forms at any level of generality. Divisions based on "essential" likenesses summarized information about a class and its subgroups in a few generalizations. To reach this goal, Gaertner and Jussieu used teleological ideas about the reasons for being and conditions for the existence of plant parts. Since all parts in any organized body had to conform with one another, some kinds of organization necessitated or excluded others. Once these traits were established, a botanist could then work out coexisting or incompatible traits.[35]

Both Gaertner and Jussieu argued that they had determined the value of plant parts on the basis of the physiology then current, but as A. J. Cain has shown, it is far more probable that they established dominant parts on the basis of observed constancy. Dominant parts were parts that in any manifold of plants classed on the basis of likenesses and differences were the last to vary. Hence Jussieu's rule that constant characters must be preferred in distinguishing the great divisions of plants, while in the subdivisions the botanist could use subordinate and variable characters.[36] Jussieu's statement of the principle of the subordination of characters and its implementation in the carpology and in the *Genera plantarum* demonstrated to the satisfaction of working botanists that the distribution of plants in a class hierarchy could attain a satisfying representation of natural affinities and that the botanist did not have to choose between his formal discipline and the natural method.

The influence of the principle of subordination did not make itself felt immediately. European naturalists were deflected from their work by the French Revolution. Moreover, application of the principle required a fresh and open mind. By the end of the century, however, botanists were again at work on the problem of natural order, and a new generation regarded the contributions of Gaertner and Jussieu as a turning point in the study of plants. "In my orders and characters," Cassel wrote in 1817, "I have chiefly followed Antoine-Laurent de Jussieu. Yet the comparative reader will find that I have given the numerical relations of the definitive characters more definitely, the characters of the nutritive organs more exhaustively, and those of the inner seed structure more correctly and completely, these latter thanks to the splendid work of the Gaertners."[37] Cassel's reservations, and those of his contemporaries, had to do with detail, and these scientists directed their efforts toward perfecting the natural method.

One illogicality, beyond the primary division based on cotyledons, was the secondary division of classes. A division on the basis of cotyledons implied a progression in nature from the simple to the complex. It seemed to follow that the primary groups in a natural method ought to contain an increasing number of secondary classes as the reproductive organs formed more combinations. "A casual glance," Lorenz Oken wrote, "shows the procedure of nature here. She divides more and more the higher she rises and the more she increases her organs."[38] Accordingly, increasing numbers in the table of organization indicated the relative superiority of the types to which these sections corresponded.

The great problem was to motivate the divisions on a physical basis and to define the resulting groups, which, Cassel wrote, "are not sufficiently determined to enter into a single arrangement . . . it is extremely difficult to find in these groups the beginning, middle, and end of formation" (ibid.). A merit in A. J. G. C. Batsch's work, "*Tabula affinitatum regni vegetabilis*, was that it undertook this thankless task and, occasionally, reached a satisfactory sequence. The exact number of subordinate groups to be arranged was regarded as somewhat arbitrary. Jussieu admitted a hundred, Batsch a hundred and twenty-five, and Augustin-Pyrame de Candolle a hundred and forty-five. As Cassel observed, "This augmentation has come about for the most part through the division of existing groups. The sequence of groups in these authors is not always the same; distant affinities are an almost unavoidable cause of these differences."[39]

There were bonds of continuity in the procedures as well as the ambitions of the German revisions of the natural method. The motivation for the use of fructification as a source of definitive characters changed. Traditionally equated with the essence of the plant, fructification was more and more justified as a source for definitive characters, because observation had established the relative constancy of the characters of the important parts of fructification. Residue groups based on negative characters, or "privation," tended to disappear as botanists succeeded in justifying their divisions with the modalities of single sets of actual characters. Gaertner's concern for symmetry, for groups of comparable generality with approximately equal numbers of members, was recognized as artificial—as an aid to study but justified only when the groups followed one another well (ibid., p. 117).

These changes in the coordination of plant groups concerned specific relations, but they also involved conceptions of the physical reality of these groups, their actual existence in nature. Scientific opinion had shifted. Even the classes and orders of plants were now thought to rest upon reasons for being and conditions of existence in certain parts, for which a basic difference in form was most readily interpreted as a physiological distinction. Group characters and plant groups, even in the upper reaches of the classificational hierarchy, were no longer considered nominal—not in botany, at any rate.

In zoology, where the forms studied were more complex and much less easily assembled, and where the traditions, tactics, and strategy of the science were different, the breach between problems of identification and the order of nature was more serious than in botany and slowed the development of a natural method.

Zoologists were less advanced in the determination of small groups and could not follow the pattern of botany, constructing new groups of animals on the basis of groups already formed. At the beginning of his career, Cuvier tells us,

> I was often struck by a vice in most of the general and partial systems of zoology; I mean the confusion or want of criticism with respect to a great number of species, and even many genera.
>
> Classes and orders not only did not conform to the intimate nature of the animals to serve conformably as a basis for a treatise in comparative anatomy; but the genera, although ordinarily better constituted, themselves offered only insufficient resources, because the species had not been ranged under each of them in conformity with their characters.[40]

Only entomology, practiced under conditions similar to those of botany, progressed in a comparable way, and even there the formation of genera was slow. And yet this approach, because of the precedent of botany and the complexity of animal form, seemed most prudent to many naturalists.

Johann Christian Polycarp Erxleben and Johann Friedrich Blumenbach, both professors at Göttingen, had been drilled in Linnaeus's *Systema naturae* by C. F. Büttner. Not surprisingly, they accepted Linnaeus's six animal classes and concentrated on distributing them in the Linnaean manner. But they proposed to overlook Linnaean characters when other strong resemblances compensated for them. Mammals, for example, exhibited many differences in form and mode of life, and little could be said of them in general. A division of mammals based upon extremities or dentition would result in unnatural groups.

In place of this artifice, Blumenbach proposed a system "based upon the total habit of these groups, but chiefly upon the instruments of movement, since these are most striking and are very appropriate to total habit." He added with respect to two orders that included a great many forms that he had "subdivided again in families according to differences in their denture and designated them with the well-known names of some of the Linnaean orders."[41]

Gottlieb Conrad Christian Storr, a professor at Tübingen, approached the problem of distribution with a procedure deriving from anatomy and physiology—and diametrically opposed to Blumenbach's total habit. On the basis of the relation between vital functions and the organs that served as their instruments, Storr proposed to evaluate the dignity of the instrument by means of the importance of the function it served. The importance of characters was proportionate to the degree to which those characters determined the animal's mode of life. As in Gaertner's carpological method, the constancy of a character provided Storr with clues to its relative importance. The regularity with which some organic dispositions appeared indicated the importance of the function they served. But Storr's *Prodromus methodi mammalium*, a direct predecessor of Cuvier's mature method, was imperfect "because he did not work on the originals." But, continued Cuvier, that method "has been perfected since, and it is in conformity with Storr's ideas that the quadrupeds, and even the whole animal kingdom, were distributed toward the end of the eighteenth century."[42]

Carl Friedrich Kielmeyer sketched a third approach to the problem of the formation and coordination of animal classes.[43] Each of Kielmeyer's vital forces—sensibility, irritability, and reproduction, to name only the most important—when considered by itself was exercised in animal organizations by unequally developed and differently constituted organs. In sensibility, for example, the capacity for manifold classes of sensations decreased as one descended the series of organizations from mankind. A naturalist could similarly differentiate and classify organizations relative to each vital force. By descending from complex to simple forms, Kielmeyer showed now a decrease in one vital force left a field free for the development of another.

He went on to argue that the force through which the developmental stages of the individual took place harmonized with the vital forces through which the series of organizations had come into being. The nature and laws of the various stages of development harmonized with the appearance of the series of forms. Sketchy and speculative as an argument and devoid of morphological detail, Kielmeyer's essay was nevertheless an influential proposal for a distribution of animals in a single continuous series.

By the end of the eighteenth century, then, naturalists in the Reich had proposed at least three different strategies to attain a natural method in zoology: the formation of families on the basis of generic affinities, the evaluation of characters by means of which the naturalist could descend from primary to subordinate divisions, and an arrangement of animals according to complexity and differentiation of organization. For all the speculative interest of these strategies, naturalists often left the question of a natural method open. The principles for a solution to the problem of natural order were not easily chosen, while the materials and tactics that would eventually lead to a solution were easily recognized.

The formation of small natural groups, although prudent, was slow. Knowledge of invertebrates was uncertain, and genera were imperfect. The approach led to error by emphasizing outward likenesses at the expense of important but obscure facts of inner organization. As a guarantee of success, then, this strategy required a thorough study of anatomy, as well as careful consideration of external conformation. The other two strategies, physiological evaluation and progressive formation, could be carried out only when anatomical data were more precise. But all three strategies were practicable once the naturalist disposed of sufficient anatomical data.

The history of comparative anatomy's gradual involvement in the animal system is known in outline, and I shall limit myself to a brief survey of the pertinent features of the technique and to brief histories of three important divisions—vertebrate and invertebrate, insects and crustacea, and molluscs and worms—established or reformed during the later eighteenth century.

Zoologists had studied the differentiation of closely related forms since the late Renaissance and had taken their characters from external traits whose functional and morphological significance they subordinated to the simpler problems of identification. In the opening phase of comparative anatomy, naturalists were divided among the many who observed only external conformation (these were usually older men) and the few who devoted themselves to the study of inner organization. Whatever their approach, though, all naturalists had to adapt their results to the conditions in which the problem of determination presented itself. Comparative anatomists, like their older colleagues, made use of class and of ordinal and generic distinctions established by their predecessors on the basis of external conformation, and they were concerned to relate inner dispositions to external marks that served as their signs.

Henri Daudin has stressed that it was this use of external marks that made it possible to arrange an unstable compromise between naturalists who did not practice dissection and the comparative anatomists. "Thanks to the notion of these 'indicative characters,' it became possible to reach an agreement between the means used by naturalists who did not practice dissection to define zoological groups and the conclusions, progressively confirmed and completed, of anatomical inquiry pursued by others."[44] Those who concentrated upon external confirmation focused upon marks related to mode of life that expressed hidden differences in organization, while zootomists concerned themselves with the concordance of inner and outer characters.

Dissection remained the basic operation in the new technique. Dissection had long been practiced on larger animals, but even after the mid-eighteenth century it had not yet been applied to invertebrates in any consistent way. The zootomist attempted, above all, to separate the various physiological systems (digestive, circulatory, nervous, etc.) and to lay out the details of their compositions, arrangements, and connections. They were given a leg up by Haller's insistence upon the comparative study of modifications of functions paralleling changes in organs. Haller had offered very fine examples of these studies in his anatomies of the eye and the brain in fish and birds. Then, in his monumental *Elementa physiologiae*, Haller assembled everything that had been written by earlier comparatists on the organs or species. His extensive citations convinced specialists of the importance of studies of this kind and stimulated anatomists and physiologists to cultivate comparative anatomy.[45]

An auxiliary technique, already brought to a high degree of perfection by Malpighi, Ruysch, and Swammerdam, consisted in injecting vessels with mercury, wax, or some other dense, liquid mass, rendering their complex embranchments visible. Beyond the material preparation of specimens, anatomical work required the use of magnifying glasses and simple microscopes. Some naturalists used magnification to observe small but complex organs but did not practice dissection under glass. Zootomists who used more complex instruments still preferred simple to compound microscopes; the aberrations of sphericity and chromatic dispersion in compound microscopes made them less desirable for observation, while simple microscopes allowed an equal degree of enlargement and produced more reliable images. The central fact about both kinds of microscope, however, was that their optical properties changed slowly. From Leeuwenhoek and Huygens to Fraunhofer and Chevalier, optical resources were limited; many organisms, organs, and tissues were beyond the means of structural investigation.[46]

However superior the procedures and technical rules of comparative anatomy were to methods that had satisfied naturalists twenty years earlier, the new discipline did not revolutionize systematic work. The comparatist continued to think of the animal type as a static form located against a logical horizon. His work was distinct from that of the field naturalist, who furnished science with new data. Field naturalists' descriptions and identifications were subject to correction by systematists located at universities and academies possessing large collections of specimens and books. Even for those unusual naturalists whose careers involved both field work and systematics—most notably Peter Simon Pallas and the Forsters, Johann Reinhold and Georg—this division of labor is striking. A much larger number of systematists were sedentary workers who, with the exception of entomologists, had seldom observed the objects of their study in a living state. Except for mammals, birds, and to a degree insects, their physiology was a mass of inference drawn from the examination of structure. Nor were naturalists interested in animal development. They conceived forms in a traditional way, and an animal in the process of development was outside their field of interest.

In short, comparative technique in the later eighteenth century was rigorous in the treatment of external conformation, precise in dissection, and effective in anatomical representation. But the discipline remained poor in physiological information, indifferent to development, and restricted, even with respect to structure, by the limitations of optical resources.

To show how three basic divisions of animals were motivated and redefined on the basis of comparative anatomy, let us now turn to some of the more generally used manuals as well as other, more specialized, treatises. The tenth edition of Linnaeus's *Systema naturae* (1758) offers a reliable point of departure.

Suppose we begin with the distinction between vertebrate and invertebrate. In Aristotelian tradition, the presence or absence of blood operated the primary division of the animal kingdom: animals were either blooded or bloodless. Many animals of the latter group contained a vital fluid that most Aristotelians regarded as analogous to blood. An alternative tradition, based on Pliny (ultimately on Plato's *Timaeus*), divided animals on the basis of habitat. Both of these traditions were alive and well throughout the eighteenth century.[47]

In the tenth edition of the *Systema naturae*, Linnaeus stated that inner structure indicated the natural division of animals, and he gave as an example a division of the six animal classes based on a number of anatomical and physiological considerations. He supplemented the Aristotelian opposition based upon presence or absence of blood and made it possible to apply that distinction in a more satisfactory way.

In addition to the consideration of the blood, in which contemporary physiology located the material factors in the development of organs, Linnaeus offered two positive characters, temperature and color. These characters, taken together, resulted in a primary division of animals into three groups: animals whose heart had two auricles and two ventricles and whose blood was warm and red (mammals and birds); animals whose heart had one auricle and one ventricle and whose blood was cold and red (amphibia and fish); and animals with one auricle and no ventricles and whose fluid was cold and white (insects and worms).[48] Linnaeus obtained the three groups of this primary division in a single operation. His distinction of vertebrate and invertebrate equalled but did not surpass the importance of warm and cold blood.

With the widespread acceptance of the Linnaean method, the idea of six animal classes acquired a firm foothold. For a time, naturalists continued to draw characters from the blood and the circulatory apparatus, although these characters led to uncertainty and indecision. In the *Handbuch der Naturgeschichte*, Blumenbach characterized the first four classes of animals solely by the color and temperature of the blood. Animals of the two final classes differed from the other classes because they had no red blood but only a whitish fluid.[49]

Although naturalists knew that all blooded animals had a spinal column, they continued to treat blood and the circulatory apparatus as the basis for the primary cut in the natural method. Contemporary physiology helped preserve this particular practice. Mechanical medicine regarded the circulation of blood as the chief agent in the manifestation of structure; the bony assemblage proper to blooded animals was secondary, a product of the circulation that led to the gradual ossification of cartilege. G. C. C. Storr, following Haller, translated this conception into the language of systematics. Because blood and the circulatory apparatus were primary agents in the animal machine, they had to be considered first in any rational classification of animals. The osseous structure proper to blooded animals, a product of the vascular system, could be accorded only secondary importance and was used to characterize blooded animals.[50]

A. J. G. C. Batsch, a professor at Jena, was among the first to break with this routine. He rejected divisions based upon the presence or absence of blood as well as upon habitat and bodily form. He took instead an old Aristotelian idea and, first, distinguished more perfect from less perfect animals. The less perfect consisted of the single Linnaean class *Vermes*. To establish the more perfect animals, Batsch opposed shelled animals (*Schaalenthiere*) to bony animals (*Knochenthiere*).[51] Although his division was imperfect, Batsch's distinction of the first four Linnaean classes on the basis of skeleton indicated the direction of learned opinion in the Reich and in France.

When naturalists concentrated on the search for affinities that would allow them to group the Linnaean classes into a smaller number of units, it was easy to see that the distinction of vertebrate and invertebrate best supported the grouping. According to Daudin, the division of vertebrate from invertebrate was first formulated, almost inadvertantly, in a memoire by the French botanist Duchesne. After dealing with the affinities of the first four animal classes of Linnaeus, Duchesne turned to insects and worms:

> Insects and worms finding themselves essentially related, I have thought to indicate this rapprochement, and since the character of their reunion is the absence of that vertebral column which descends vertically from the summit of the head to the base of the trunk and by means of which there is established throughout the body an intimate communication with the brain, it has seemed natural to indicate this negative character with a word derived from the vertebrates: *vertebrosa animalia* signifies provided with vertebrae; *invertebrosa*, then, designates those lacking vertebrae.[52]

Lamarck and Cuvier gave the division a definitive formulation. Lamarck considered the division one of the primary pieces of evidence for a "degradation of organization from one extremity of the chain of animals to the other." Cuvier retained the Linnaean division for some years, but in the first volume of the *Leçons d'anatomie comparée* he adopted Lamarck's character. In the same volume, Cuvier tied the taxonomic unity of vertebrates to a common plan of structure, a conception that then became an integral part of comparative anatomy.[53]

Imperfect and fragmentary as knowledge of the classes and orders of blooded animals was in the later eighteenth century, the situation with respect to bloodless animals was worse. Only insects had been studied carefully, and their table of organization was enormous. As for the other bloodless animals, observation of living forms was rare, and anatomical studies were inadequate. Linnaeus had tried to establish some order in this corner of the animal kingdom, but his inadequate information forced him to apply arbitrary measures, with strange results.

Linnaeus began by simplifying Aristotelian tradition. In place of Aristotle's four main genera—*Mollusca*, *Malacostraca*, *Ostracodermi*, and *Insecta*—Linnaeus substituted two classes—*Insecta* and *Vermes*. His definition of insects was broad: it included those segmented animals whose body was covered with an osseous crust and with antennae fixed to the head. The class included the Aristotelian *Crustacea* but excluded the previously included worms. The Linnaean class was approximately equivalent to the modern phylum of *Arthropoda*. In the tenth edition of the *Systema naturae*, where Linnaeus based classification upon internal structure, he added another character common to all the animals of this class—lateral pores for respiration.[54]

Linnaeus subdivided insects into seven orders, the first six of which he based upon differences in the conformation of the wings. Here the system is, said Nils von Hofsten, "a scientific achievement of a high order . . . which in its essential parts has stood for two hundred years."[55] Under the label *Aptera*, however, Linnaeus collected all of those animals that met his class character but that had no wings: parasitic insects, aquatic crustacea, arachnids, isopods, and myriapods.[56] Taken as a group, Linnaeus's seven orders of insects had some undeniable affinities. "The first distinctive character of insects," Blumenbach wrote,

> is their singular conformation; head, corselet, and abdomen are separated from one another by incisions. . . . In the second place, these animals are distinguished by very irritable particular organs that they bear on their head when they have undergone all of their metamorphoses: these are antennae. . . . Finally, insects have articulated feet . . . and the number of these feet is always greater than in animals of other classes.

But the *Aptera*, Blumenbach continued, "are insects absolutely without wings; they vary extremely with regard to size, conformation, resort, nourishment, organs of manducation, the numbers and size of the feet, eyes, etc. Some are viviparous, others oviparous; the flea excepted, they do not undergo metamorphosis."[57] Specialists did not consider these characters adequate to define insects as a single class. By the end of the eighteenth century, the Linnaean *Aptera* constituted a separate class, and entomologists had established a superior unit, reconstituting the old Linnaean class at a higher systematic level.[58]

Charles de Geer, one of Linnaeus's younger Swedish contemporaries, took the first step in this reform. De Geer considered the *Aptera* insects, but he established them as a general group, opposed to winged insects. He divided the new group into two orders: animals that do pass or do not pass through metamorphoses in their development. The first of these orders contained a single group that contained a single genus: the flea. De Geer divided the second order, insects that do not transform themselves, into three groups based upon the number of feet and upon the presence or absence of a marked separation between the head and the corselet. The first of the resulting groups included hexapods without wings and larval acarids. The second group included adult acarids, arachnids, and crustacea having no more than eight feet. The third group included both the crustacea with more than ten feet and the myriapods. De Geer's use of single external characters led to some strange combinations and divisions, but he did manage to emphasize the special character of the *Aptera*.[59]

At this point, Johann Christian Fabricius proposed a method analogous to the division of quadrupeds on the basis of denture and of birds on the form of the beak. He divided insects according to their buccal apparatus. Fabricius found the materials for his analysis in works by Swammerdam, Réaumur, and de Geer. He himself dissected the mouths of all insects that came into his hands. Eliminating the old division between the winged and wingless, Fabricius separated hexapods from crustacea, arachnids, and myriapods on the basis of masticative apparatus. The division, although exact, was impractical for a working naturalist; the parts from which Fabricius took his characters, said Cuvier, were too small for identification of genera, and the exclusive use of buccal apparatus was artificial and led to results that disturbed the feeling many naturalists had acquired for natural affinities.[60]

While a student at the Carlschule in Stuttgart, Georges Cuvier had studied the methods of Linnaeus, de Geer, and Fabricius, and early in the next decade he discovered an important organizational difference: the presence of heart and branchia in crustacea and the absence of these organs in insects. Cuvier used this character as a basis for two groups. The location of a respiratory system in a single region of the body and a propulsive organ of circulation as its complement indicated a mode of life in crustacea different from that of insects and more closely allied to that of molluscs and vertebrates. Cuvier's distinction proved to be a reform of classification and a definitive statement in the comparative analysis of vital functions.[61]

The naturalists who studied *Vermes* were numerous, and I limit myself here to those who busied themselves with molluscs.

Linnaeus used *Vermes* as an all-purpose class: it included all those bloodless animals excluded by the positive characters of *Insecta*. This miscellany included worms but also the conch, nautilus, starfish, medusa, and sepia. The division of this class in the first edition of the *Systema Naturae* into *Reptilia*, *Zoophyta*, and *Testacea* was determined by the needs of identification. Linnaeus based each of these orders on a few simple, distinct marks.[62] When Linnaeus reformed this classification in the tenth edition of the *Systema naturae*, he tried to define *Vermes* with common internal characters, but the marks attributable to the class were either purely negative or inexact and indecisive. Linnaeus added two new orders, *Lithophyta* and *Zoophyta*. The order previously called *Zoophyta* now became *Mollusca*. Four orders of the class—*Intestina* (formerly *Reptilia*), *Testacea*, *Lithophyta*, and *Zoophyta*—were given distinct external characters and contained genera that for the most part corresponded with the new ordinal characters.

But *Mollusca* became another all-purpose category. Of its six subdivisions, only two could in any sense be considered natural. The best-known example of systematic caprice is Linnaeus's separation of shell and animal in the table of organization. He relegated the common limpet, *Patella*, to the *Testacea*, but the animal in the limpet was a snail, *Limax*, and belonged to the *Mollusca*. The animal in the snail *Chiton* Linnaeus classified as the sea snail *Doris*, which found a place among the naked molluscs. There are similar examples of the Linnaean *esprit de système*. *Actinia* and corals were divided among *Mollusca*, *Lithophyta*, and *Zoophyta*. Worms were found in all five of his orders, while the slime eel, *Myxine*, was found among the true worms.[63]

Specialists regarded these arrangements as unworkable. Johann Baptiste Bohadsch of Prague, for example, who was studying marine animals in Naples, grew impatient with Linnaean reform, and his treatise became an extended commentary on the inadequacies of the succeeding editions of the *Systema naturae*.[64] Blumenbach acknowledged that *Vermes* had no common character. Worms, he said, "have so few positive and general characters, that the shortest way to define them is to call them white-blooded animals that are not insects, that is, that differ from the latter by the absence of antennae and articulated organs of movement."[65] Blumenbach retained the Linnaean class, and his subdivision, like that of Linnaeus, was regulated by the need for easy identification.

The principle for a projected reform of invertebrates was simple but required a break with tradition. Among those studying invertebrates, only Pallas had stated the principle clearly and applied it consistently. Neither the Linnaean *Vermes*, which Pallas called *Mollusca*, nor its orders were well founded. *Testacea*, for example, taken from conchyliology, was based upon the secondary character of the shell. If, however, the anatomist examined not the shells but the animals inside, he would find three families of *Mollusca*, all of which included animals with and without shells.[66]

Georges Cuvier took up and developed Pallas's proposal thirty years later in two memoires on the structures and affinities of worms. Cuvier rejected amateur demands for easy identification: if beginners needed a division in order to find the names of species, let them use an artificial system. In the "grand système des Mollusques," shells formed only a particular system. "My aim has been to make known more exactly the nature and the true relations of animals with white blood by reducing to general principles what we know of their structure and of their properties."[67]

To this point, I have attempted to reconstruct the chronology for three of the superior zoological divisions reformed or established in the opening phase of the implementation of comparative anatomy. This anatomical reform involved a profound change in the conception of the natural reality of these divisions. The first cut in the animal kingdom, as well as the immediately subordinate class concepts, were no longer regarded as nominal or utilitarian. The distinction between vertebrate and invertebrate, between insects and crustaceans, and among the orders of *Vermes* were reestablished between 1760 and 1790 upon actual physical differences, which naturalists regarded as both essential and constant. And for comparable groups, from cetaceans to salamanders, naturalists were well on the way to applying equivalent general principles to positive physical apparatuses in order to produce, by means of a hierarchical arrangement, a satisfactory representation of natural order. In this complicated process, abstract groups based upon privation were dissolved and then reformed on positive physical bases. *Insecta Aptera* is an obvious example. The traditional concern for equal numbers of members in groups of equivalent rank was recognized as artificial. De Geer, for example, did not hesitate to give fleas an order and genus of their own.

The decisive factor in this change was purely technical: the implementation of anatomical comparison. The reformed study of organization, which made it possible to analyze internal and external dispositions, slowly put an end to equivocation with regard to defining characters. Once naturalists had begun to identify the full extent of the significant concordances among animals, they could replace the impossible project of a method that expressed all natural affinities with rational principles that working naturalists considered as certain as the data themselves.

The source of these principles lay in the conceptual tradition of natural history. Naturalists continued to group animals on the basis of common characters and to construct class concepts on the basis of those characters. Traditionally, common characters were not arbitrary collections but essential nuclei that determined the structures of all group members. Comparative anatomy, for its part, regarded structure as governed by mode of life and the various physiological systems that constituted structure as determined by function. Parts whose conformation and activity were in direct relation to principal modes of activity were thought to determine the functional organization of the entire animal. The presence of a heart and branchia in crustacea, for example, indicated a mode of life necessarily different from that of insects.

The combination of these two ideas (the relative value of characters and the physiological determination of organization) led, by the circuitous route of which I have described only the very smallest part, to the principle of the subordination of characters and to the possibility of conciliating natural order with systematic dichotomy. From being inductive, or rather intuitive, classification thus became, at least at the more elevated levels, deductive or constructive.[68]

Different as the materials, methods, and achievements in botany and zoology were between 1760 and 1790, the two sciences were united by a common project: the achievement of a natural method and, by a fundamental procedure, the so-called subordination of characters.

The principle of subordination determined the relative importance of organic parts in two ways. Naturalists calculated the dignity or perfection of a part according to a priori notions, and I have examined some of their attempts to weigh characters on the basis of the gravity of the function served. Traits of conformation that exercized the most marked influence over the organized body were called *dominant characters*. Other characters were subordinate to these, and there were various degrees of subordination and, hence, of functional importance.

Along with this a priori calculation, the naturalist employed simple observation. Dominant characters in natural groups were constant. In any manifold of natural objects brought together on the basis of likenesses, the dominant characters were the last to vary. From their constancy and influence, equally, resulted the principle: dominant characters were to be preferred for primary divisions of the natural method. As the naturalist descended to subdivisions, he also descended to subordinate and variable characters.

Many taxonomists will be familiar with A. J. Cain's demolition of this procedure. Naturalists observed the constancy of characters in groups formed before any particular character was checked, and much of the a priori approach actually rested upon the observation of what parts and systems of parts varied least in groups already constructed.[69]

In German work before 1790 we find the rudiments of what was to become in the hands of Georges Cuvier the principle of subordination. I have just considered a very small part of its use in the animal kingdom in the establishment of vertebrates, crustacea, and molluscs. In the analogous case of plants, practice was equally hesitant, as we have seen in the case of Joseph Gaertner. It was only in the work of the early nineteenth century that the subordination of characters became a generally utilized procedure in natural history everywhere in Europe. Naturalists, in realizing the natural method, confronted one great problem—the conciliation of their formal discipline with natural order. The use of subordination as we find it in Gaertner's carpology must have seemed a great step toward the solution of the problem. The division of forms at any one level in the class hierarchy rested upon the distinctive traits of one part; but the choice of characters at every level also conformed with all other relations. The difference between the vital importance of parts and the classificational value of characters was covered by an a priori physiological rationale.

In their daily work, naturalists disposed plants and animals in series, whose members related to one another by a greater or lesser number of affinities. It was assumed that these series, when complete, would exhibit a seamless continuity from one end to the other. A less pressing concern was to justify this state of affairs, to establish the principle of continuity in such a way that it harmonized with other elements of science—the fixity of species, for example, and the economy of nature. Only after a great deal of information about likenesses and differences had accumulated was it clear that the traditional image of a single ascending series of forms was inadequate to represent the affinities and analogies in play among natural forms. The discussion of coordination continued to center upon the notion of continuity.

In 1745, Vitaliano Donati had compared natural affinities to meshes in a net. Natural order was "a tissue of many threads that communicate, relate, and unite with one another."[70] Johann Hermann, in his *Tabulae affinitatum animalium*, gave the net image its most elaborate realization: affinities, he said, dispersed and expanded laterally; at other points, they contacted one another and related mutually.[71] On the printed page, Hermann drew up a table in three sections, in which Linnaean classes, orders, and some genera and species were connected by lines that symbolized important resemblances. Hermann argued that these relations had a degree of natural affinity, but of course no two-dimensional table could simultaneously do justice to the unequal degrees of perfection. Eventually, Lorenz Oken brought together the best features of the net image. The schema that best represented animal affinities was "neither series, nor net, but a stereometric net, a series whose basis is a net."[72]

In an early work, *Elenchus zoophytorum*, Pallas argued that continuity bridged all cleavages between living forms. "Nature," he wrote, "having begun with the simplest things, changes and forms organs little by little in relation to their ends; on a single construction, she grafts, she adjusts successively new parts."[73] The passage parallels Gaertner's discovery that, as organic combinations grow more complex, they establish affinities in all directions. Neither Gaertner nor Pallas could find anything in this fact to justify the traditional image of a single gradation. Pallas proposed instead the image of a tree. The root, representing the simplest natural forms, issued in two trunks: plants and animals. The trunks were separate, but the branches established lateral relations (ibid., pp. 47–48). Pallas did not offer this image as a genetic representation of continuity; he simply wished to give the clearest possible synchronic image of the multiple affinities among living forms.

In the same work, Pallas also proposed a stereometric arrangement. After criticizing Bonnet's series, he added: "Not less good, if not far better, would be to imagine the families of organic bodies as distributed next to one another so that their many surfaces and subdivisions represent a polygon in which the various levels of their affinities could be made clear" (p. 47).

Other images of continuity were current in the later eighteenth century, each of them designed to deal with void or dispersion in various ways. I have mentioned Linnaeus's geographical map in botany, which was adopted not only by Gaertner and Jussieu but by Bernardin de Saint Pierre, l'Heretier, Petit Thouars, and de Candolle. A map, says Giulio Barsanti, is an appropriate image when we discover that nature is far more complex than has ever been supposed: the map implies that multiple affinities, differentiated and intertwined, make grouping in bunches obligatory.[74]

The alternative images of continuity, all spatial, all synchronic, attempted to capture the complicated relations now seen to hold among natural groups. If each group were plotted in space according to its observed affinities, the result would be neither simple linearity nor random confusion. The groups would be distributed as the knots in a net, the branches of a tree, the facets of a crystal, or the territories on a map, that is, in an irregular but nonetheless continuous form. The relation between these images and classifications of plants and animals was complex. All of these images appeared in introductions to works of classification as parts of extensive programmatic statements. Barsanti argues the priority of classification over image. The images of ladder, net, and map may have reflected the classification, and naturalists may even have attributed a constructive role to them in their distributions, but in fact, says Barsanti, the effect of increasingly sophisticated classifications was to weaken preexisting images to the point of destroying them, as in Pallas's rejection of Bonnet's graduated series (ibid., pp. 98–99).

But the situation could be even more complicated than Barsanti suggests. Let me borrow a single case from an unpublished paper by Peter Stevens. Throughout his career, Jussieu held to the idea that nature formed an unbroken series. He did not, however, believe in an unbranched continuity. In adapting Linnaeus's idea of a geographical map, Jussieu emphasized his particular feeling for continuity. Between the boundaries of different administrative units on a map, there were no physical gaps, and in one step one could move from one province to another. This conception, when applied to the coordination of plant groups, circumscribed the possible affinities that plant groups could manifest. Species, genera, and orders were coterminous with the species, genera, and orders that surrounded them. And this particular conception of continuity pervaded the coordination of plant groups in Jussieu's natural method, as Stevens has shown in convincing detail.[75]

Blumenbach's well-known criticism of these images raises most of the underlying issues with which he and his contemporaries were struggling. The beloved images of a chain, a series, a net, and so on had an undeniable heuristic application, and their value for the natural method was not to be denied. But, Blumenbach continued, "they do not have so real a basis in nature that we must attribute them to the Creator in the plan of creation;

and the perfection and continuity of nature cannot be sought in the fact that, as one expresses it, Nature makes no leaps, since the creatures with respect to external habit can be arranged so neatly in a series."[76]

But if natural order was not grounded in the will of the creator, what was its real basis? Was it, as Blumenbach suggested, echoing Kant, merely a regulative, methodological principle? The answer seems obvious. If natural order was not grounded metaphysically, then it must be a product of physical and temporal processes. Blumenbach was aware of this alternative, but he found the prospect impossible and so opted for order as a regulative, interim principle.

Naturalists trained in orthodox systematics—Blumenbach and Georg Forster, for example—were not slow to point out that the genealogical method advocated by Buffon (and followed by Immanuel Kant) had no purchase above the species level and was wholly impractical when applied to almost all animals. Blumenbach and Forster quite consistently continued to use common external characters to define natural groups and their arrangement.

Among naturalists in the Reich, the earliest and for a long time the only positive reaction to ideas on physical continuity was found in the works of Pallas. The preface to the *Elenchus zoophytorum* shrewdly analyzed the concepts of affinity and continuity. Many obvious examples, Pallas wrote, confirm that nature makes no leaps. "She has caused the whole host of organic bodies to succeed one another in the greatest order; and according to the most careful sequence of affinities, she has arranged the species in genera, these in orders, the orders in classes, and the classes again among themselves."[77]

But this sequence of affinities had nothing to do with the "superficial and ideal affinities" used in the constructions of Bonnet and Bradley; nature, Pallas said, preferred to observe "structure, decreasing body size (*abstractiori habitu*), and mode of reproduction." Nature began with the simplest things, changing and forming organs little by little in relation to their ends: "on a single construction she grafts, she adjusts successively new parts." And this was the real source, the physical and temporal source, of continuity. "From this the great confirmation we derive from the likeness of bodies (and which we must not use without mature consideration) takes its source (ibid., pp. 46–47).

This remarkable passage follows lines laid down by Buffon in his conception of physical continuity. Pallas's distinction between, on the one hand, superficial and ideal affinities based on analogy and, on the other hand, natural affinities based on structure and reproduction builds upon Buffon's division between the abstract order of ideas and the real order of bodies. Unlike Buffon, Pallas had firsthand knowledge of the forms he was classifying and a firm grasp of the questions posed by his fellow naturalists. And for these reasons, his discussion moved beyond Buffon's abstractions in at least two important ways.

Pallas saw that the term *affinity* blurred form and function. A bat's wing and the lateral fins of a flying fish, to appropriate Pallas's example, served movement in analogous ways, but in structure they were totally different. Conversely, structurally similar organs might exhibit modifications and perform very dissimilar functions. All of this was confused in the term *affinity*, and it was uncertain whether an affinity indicated the relation between forms or in fact constituted it. Pallas did not propose a new term, but he did state that the affinities used by nature were physical and implied that only characters derived from these should constitute natural groups.

Pallas's conception of the coordination of natural groups was equally physical. Nature began with the simplest forms and gradually altered organs in relation to their ends, grafting and adjusting new parts. A natural method, using likenesses among natural bodies, would seek to reflect this physical continuity. On the other hand, a naturalist would avoid the superficial and ideal analogies that either juxtaposed bats and birds or seized upon the newly discovered *Ornithorhyncus paradoxus* as an all-purpose intermediary form.

An obvious trait of the sciences of natural history in the late eighteenth century was the continuing concern for the rules and forms of classification. European naturalists who dealt in problems of identification, often on a worldwide scale, needed a precise and dependable instrument for relating natural specimens to information about properties and uses. These men found the practical part of botanical and zoological classification, the method for establishing the names of plants and animals, indispensable for their daily work. The rules of nomenclature, the choice of characters, the number and order of the categories superior to the genera—all of these were common concerns among European naturalists.

But even after they had persuaded one another that the pattern of groups within groups produced by their formal method was, potentially, an accurate representation of natural order, these men did not equate that pattern with the idea of common origin. They did not do so because their conception of natural order continued to rest upon assumptions and modes of observation different from those that led to ideas of common origin and descent.

The continuity of nature, even in the modified versions current in the late eighteenth century, was a necessary trait of the rational order inherent in the relative perfections of natural forms and possessed its own reasons for being. In principle, a system representing the relations of organized bodies could be established without reference to genealogy or history. Natural order, the affinities and analogies among the forms that constituted it, and the harmony and balance of their disposition, far from requiring a genealogical table for its clarification, rendered the project inconceivable or useless from the naturalist's point of view.

One can cite many statements from the closing years of the eighteenth century that point to a quickened interest in viewing natural forms in terms of their history. But the connection between these programmatic statements and the practice of naturalists was tenuous. The fundamental codes of natural history—the assumptions, practices, techniques, and modes of reasoning that crystallized in the terminology, descriptive models, and armature of class logic—were not subject to discussion and revision in quite the same way as the interpretations that explained the reasons for and the causes of natural order. On the basis of this practice, naturalists delimited many classes and orders of plants and animals with a precision confirmed only by later research.

Naturalists' statements of program and principles, on the other hand, were often vague and indecisive. Rigorous systematists settled for statements to which they did not attribute any decisive importance. The guarantee of the reality of a group of plants or animals was provided by observations proper to that particular case, the "affinities" that united the members of a group more closely to one another than to other groups. The intuition that founded a naturalist's belief in the existence of a group was, as Daudin has pointed out, the soul of classification, but the precise nature of the unity remained undefined.[78]

And what of those naturalists—Peter Simon Pallas, for example—who were not satisfied with this state of affairs and sought a more profound connection between the unity of forms and the explanatory formulas? Pallas, it is clear, was preoccupied from the beginning of his career by the dual axes along which the objects of his study were located: the axis of simultaneity, where his contemporaries charted relations among coexisting forms, and the axis of succession, where, as Buffon had suggested, the repetitions and variations of forms were located. The problem Pallas struggled with was how to establish some equivalence between the different concerns of his two ordering principles.

III

The Scale of Diversity

Systems of nature and natural methods are among the most characteristic and enduring achievements of the eighteenth century. All of these compendia, from the most circumspect regional flora to the most ambitious inventory of the terraqueous globe, are inspired by the belief that nature's own plan is not only fixed but known, at least in outline. This enormous production, which focused upon the constant, observable aspects of natural order, has always overshadowed another, smaller, but equally important literature written by and for specialists. The transactions and memoires of academies and universities of the second half of the eighteenth century are full of studies of degeneration, hybridization, metamorphosis, monstrosity, and race—the irrational backside of order and stability.

Naturalists and physiologists recognized these kinds of variation with reluctance and tried to limit their effect on the system of fixed natural forms. As knowledge about any one kind of divergence from established order accumulated, specialists segregated the subject and treated it as a separate set of problems. Their separate studies did not constitute an integrated body of knowledge, not even at the end of the century, but they did move research in a new direction, and they were inspired by a common belief. Naturalists were forced to confront the physical processes constitutive of living forms and to invent new methods for the study of these processes. Even though some kinds of diversity seemed to threaten to dissolve the basic units of natural order, that is, the species, it was still possible to believe that these kinds of diversity could be made intelligible and, moreover, that they might even throw light upon the world of stable, established order.

Working naturalists, who dealt on a daily basis in problems of identification and characterization, mastered the practical problems of specific diversity by supposing two different natures united in each individual form. One set of traits was constant, the same in all members of one species. Another set of traits was variable, changing from individual to individual according to physical circumstances. The two natures, combined in one individual, were nevertheless distinct. The individual proper was a local ephemeral fact; the representative was a constant element in the system of nature. For many, natural history was the science of representative types studied through individual forms. A complete inventory of these types was considered by some to be the basis of all science at the species level. Although this was the accepted strategy for dealing with specific diversity around 1750, it did not solve all problems, and it did not pass unchallenged.

Of the conceptual schemata opposing the fixed forms of essentialist systematics, none was more widely read and commented upon than that of Buffon. As we have seen, Buffon, in the first volume of his *Histoire naturelle*, had substituted physical descent for formal criteria in defining species and had sketched a vague genealogical method based upon prototypes. Nature had employed a single idea in the creation of animals and over time, by means of perfection and degeneration, had produced the races and kinds of animals now known.[1] The ideas of time and descent that were implied in Buffon's conception of the prototype (rather vaguely, to be sure) offered an alternative, physical interpretation of diversity, which was attractive to some contemporaries. Working naturalists, however, resisted defining species in terms of reproductive criteria, and they were puzzled by the idea that species were transformed over time.

The essential likenesses of species members, including their tendency to breed true, did not exclude partial differences, of course. Accidental, secondary characters complemented specific constancy and offered a point of departure for the varietal concept. But among secondary differences, where did a variety begin and end? Did the varietal concept include the slightest and the greatest variations, the most fugitive and the most constant? Among the differences from individual to individual, were not some differences greater and some lesser than the variety?

Linnaeus defined the varietal concept broadly: "Varietas est planta mutata a causa accidentali."[2] Accidental causes were external influences—climate, soil, temperature, and so on. Any change produced in the original species by an accidental cause, any deviation from the type, was a variety.

The broad definition did not last, not even in Linnaeus's own work. Practice worked against any inclusive concept. Gradually, usage was situated somewhere between the slight nuances that gave each natural form a physiognomy of its own and the grave deviations that altered the type profoundly. Naturalists also learned to distinguish the varietal concept temporally: the concept applied to a greater or a lesser number of individuals but was not considered common to all forms descended from a common ancestor. Constant deviations constituted a subtype, the race. The difference between a variety and a race consisted in the relatively greater constancy of the race perpetuated by generation. Kant later added a further measure of constancy: when two races crossed, each necessarily contributed its character to the progeny.

Terminology in this area flourished. Eventually it included not only *Racen* but *Spielarten* and *Schläge*, each with a special set of criteria. Naturalists continued to attach two distinctions to the term *race*, however: community of origin and hereditary transmission of variation. A race was a class or series of individuals issued from one another and distinguished by a variation that had become constant. Naturalists considered these subtypes distinct branches, in spite of their common origin, and recognized that yet further subtypes might arise from them.

Unlike the intraspecific crosses that produced races, crosses between species were thought to produce sterile offspring or offspring whose fertility was limited to a few generations. The sterility was a consequence of the fact that these unions were outside the rules of nature. The possibility of a fertile union between beings whose natures were essentially different might be admitted, but the latitude accorded to these unions was not known. Linnaeus had convinced himself of the possibility of crosses between classes and orders of plants; and from various discussions of Buffon, Réaumur, Haller, and Bonnet, it is clear that other naturalists believed that crosses between very different organisms were possible among animals as well. The opposing tendency in research was to restrict the field of hybrid generation to the smallest number of cases, usually domestic.

Naturalists also studied the degrees of resemblance between parents and hybrid offspring. Did hybrids necessarily resemble the mother rather than the father, or were they rather mean products? The question of sterility in hybrid offspring was as old as natural history and was not unmixed with religious ideas. Everyone conceded that products of two races or two varieties were fertile and that following generations were equally so. On the other hand, products of two individuals of different species were often infertile or, if fertile, were so for only a few generations. Hybrid fertility, it was thought, was noncontinuous.

As the term indicates, *hybrids* (or *bastards*) were considered anomalous; they resulted from a union outside the limits prescribed for species. Other anomalies, ranging from simple nuances to complex vices of conformation, exceeded the prescribed limits in their modifications of organization. But the very possibility of classifying monsters according to excess, lack, reversal, or false position of parts indicated the regularity of abnormal organizations, whose most obvious character was, according to Blumenbach, the striking uniformity of their formation, a circumstance obviously related to determinate laws.[3] These, when added to the laws that regulated normal organization, would unite teratology with natural history in a community of principle.

The kinds of diversity I have been discussing rested upon one obvious fixity: the representative type was conceived by means of an indefinite succession of individuals, and the likeness was that of an adult, the definitive form. But naturalists had long since established that individuals of a type or subtype were often divided between two distinct states (the two sexes, for example), or among three or more distinct states (drones, queens, and workers, for example), or between different generations. Moreover, the notion of metamorphosis, according to which living forms moved through distinct phases, underwent a revision in this period and was established as one of the most general facts in life cycles. Many forms traversed a sequence of phases toward either a single distinct state or two or more distinct states. In these cases, the type could not be represented by a single individual or even a couple but could be represented only by a group and, in some instances, by a cycle of generations. The species was a class of individuals that were not all alike yet were related by manifold distinctive characters scattered among different members.

The improvisational character of the study of variation is clear. An even more striking character is the correlation between structural similarity and common origin. The formal likenesses of the essentialist species concept, for example, involved reproduction as a means of perpetuating the form. Various kinds of diversity—races, hybrids, and monsters—could be explained only by an alteration in the normal principles of reproduction. In fact, all eighteenth-century discourse on the species and on specific diversity involved a theory of generation. At its simplest, the theory was reduced to the idea of succession, an unbroken chain of existences back to either creation or a first progenitor. The different kinds of diversity involved the additional consideration of individuating conditions and principles acting in time and opposing universally valid general laws. It is at the level of specific types and diversities that the notion of viewing organized bodies in terms of their genealogy, rather than in terms of atemporal affinities, acquired a foothold in natural history.

A point of entry should not, however, be confused with the triumph of a historicist point of view. We shall see, in fact, that because history was involved it suggested that variation was fundamentally of no importance. The individual, mostly unrelated, problems associated with succession, change, and diversification did not add up to a transformation of the criteria by which naturalists established the species, nor were reproductive criteria extended above the species level. At the very most, I think one can speak of an intensification of interest toward the end of the eighteenth century in viewing natural forms in terms of their history. But this is an argument that can be defended only after we have examined a representative selection of research on organic diversity.

The principle that variation was a consequence of external physical conditions was supported by many facts, and when we confine our attention to plants, it is easy to sum up received opinion. Temperature produced many variations. In a cold climate, plants grew smaller and frailer than ordinarily; their leaves were often caducous, and their fruits aborted frequently. The same plants in a warm climate were larger, woodier, and more luxuriant; their colors were livelier, their savors pronounced.

The effects of light and water were equally obvious. With inadequate light, plants were pale, fragile, and etiolate; in direct sunlight, they grew stronger, tougher, and more highly colored and were often hairy. In a single climate, humidity produced many variations. Plants growing in water lost their hair, altered their leaves, and lengthened their stems in an attempt to reach the surface of the water. When, on the other hand, plants accustomed to humidity were forced to live in drier terrain, they covered themselves with hair, grew smaller than ordinarily, and acquired a degree of toughness.

The influence of soil was no less obvious. In compact soil, roots were small and contracted, while in sandy soil they were large, fleshy, and highly developed. Where the soil contained salt, leaves were often fleshy and glaucous. These different factors might be combined with one another in nature to multiply the causes of variation; cultivation could augment any effect by further diversifying circumstances.

These kinds of variation were among the most thoroughly studied phenomena in eighteenth-century botany and horticulture. Between the publication of Linnaeus's *Flora lapponica* and Alexander von Humboldt's *Essai sur la géographie des plantes*, an increasing number of books and memoires studied the relation between variation and the physical nature of the locality in which plants grew.[4] Naturalists concluded that variations due to climate and nourishment often affected only individuals, altered only color, size, or proportion, and lasted only as long as the causes that produced them. In other words, varietal differences were anomalous, superficial, and ephemeral characters playing around a common center, and the principle that variations were caused by external agents was regarded as one of the strongest supports for the doctrine of unchanging species.[5]

Another important source of variation showed, however, that not all anomalies were determined by external agents. Believing that plants of different species could, under some circumstances, fertilize one another in such a way as to produce new kinds, Linnaeus had proposed that in the beginning there were just as many different plants as there were natural orders. These plants had intermixed to produce new genera, and these in turn by reciprocal marriages had produced all the existing species and varieties.[6]

Between 1760 and 1780, however, the study of crosses and crossing had progressed sufficiently so that Peter Simon Pallas could minimize the importance of hybridization as a source of new species. "During more than a century during which we have observed plants and animals with particular care, we have not been able to prove that a single distinct and constant species in nature is of modern origin."[7] However plausible Linnaeus's hypothesis, especially for genera in certain parts of the world where nature had varied the species copiously, Pallas argued, crosses between species were difficult to produce and impossible to perpetuate.

In genera abundant in species, naturalists often found two, three, or more species made according to one model yet constantly distinct among themselves. In classes and orders where varieties were rare, crossing was impracticable. The resemblances that linked species, genera, and even natural families endured, without, however, producing new links by means of the operation of nature alone. Finally, fossils that went back to the first age of nature exhibited the same varieties still found in Europe. Pallas concluded that it was probable that similar species, or species that linked genera, were primitive, "projected in the first plan of creation and destined to form that chain of beings that we admire without being able to account for it."[8]

As Pallas intended, this evidence reduced the significance of crosses and crossing as a source of new species. But this kind of multiplication retained its importance in the formation of varieties. The varieties of a single species produced by climate or nourishment, when brought together, were able to fertilize one another and to give birth to intermediate strains, varieties of a secondary order that were more or less constant. This kind of variation was common in orchards and in florists' gardens, where it was an important means of varying fruits and flowers.[9]

Among cultivated plants, the crossing of strains was the most frequent cause of variation, and each of these secondary varieties, placed in diverse stations, might develop further variations, so that the number of forms possible in some species seemed infinite. The sheer number of these variations confused those botanists who were concerned above all to establish the lines of demarcation among species. Often very disparate varieties whose origins or intermediaries were unknown passed for species; equally often, essentially distinct species were confused with varieties.[10]

The permanence of the different kinds of variation also showed itself to be very diverse. There were, first of all, purely local varieties. A plant grown in fertile soil or full sun was larger in some or all of its parts, but when shoots or seeds were planted in less favorable circumstances, the parts resumed their ordinary size. Knowledge of this kind of variation was useful to a gardener but had little systematic interest. A second sort of variety, somewhat more important to science, was also produced by external agents, but once it had been created, it remained permanent in the individual and could be indefinitely extended by means of grafts and shoots, although it could not be preserved by seed. A third sort of variety, ignored as long as possible by systematists, was preserved through seed. Gardeners knew that variations of *Digitalis purpurea* and *Lychnis chalcedonica* with white flowers could be passed on through seed and that nuances in peas, beans, and generally in all cultivated annuals were preserved through seed. Gardeners customarily used multiplication by seed to produce the greatest possible number of varieties, while naturalists employed the same means to establish true species.[11]

Linnaeus advised botanists not to concern themselves with cultivated plants; he limited his varietal concept to aberrations of the original form not propagated by seed or tending to disappear with any change of culture. But his rule was not without exceptions; Linnaeus was himself aware of variations of true species, which he believed were the effect of soil, climate, or other circumstances but which perpetuated themselves through seed in the state of nature. He had occasionally persuaded himself that these were true species. His *Veronica multifida* was only a variety of *V. austriaca*, and *V. virginica* was perhaps a variety of *V. sibirica*.[12]

In other cases, where the varietal nature of the plant was unmistakable yet preserved through seed, some botanists believed that these deviations, on the analogy with cultivated varieties, would eventually prove more or less tenacious. How many generations would suffice, de Candolle asked, "to affirm that two given plants are species or varieties? It is this that is impossible to affirm, and it is relative to these cases, happily quite rare, that doubt must exist on the demarcation of species."[13] The doubt, he added, was even more restricted once the naturalist had mastered the true distinctions of specific and varietal characters.

This is a faithful account of the problem of variation, from Koelreuter's *Vorläufige Nachricht* to Willdenow's *Grundriss der Kräuterkunde* and Kurt Sprengel's *Anleitung zur Kenntniss der Gewächse*.[14] Botanists assumed that original species not only continued to exist but were unaltered in their primitive type. These types could, of course, be altered superficially through the action of external agents and the crossing of races; the changes produced by external agents were in general modifiable by contrary circumstances, while those determined by heredity could be modified by further crossing of races.

Naturalists who worked with plants daily, and whose knowledge often included thousands of kinds, were content to repeat the worn phrases, that variation was wholly a consequence of external influences and that true species preserved their essential characters intact. They offered many facts in support of the principle that variation in plants resulted from external influences, but we should be very much mistaken if we believed that they had been led to these facts by observation alone. On the contrary, the principle was a consequence of their assumptions about the fixity of species, subsequently supported by many observations. Of course, no one would deny that external circumstances could cause modifications in plants, but for the most part these had little permanence and were therefore considered of slight systematic importance. It was the more or less constant varieties that threatened the facile equation between variation and external agents.

Botanists were aware of two series of facts that showed that variations were not due entirely to the influence of soil and climate but were partly determined by inner causes. In many cases several different varieties of a single species were found growing at one station in the same external circumstances, just as gardeners cultivated various races of one species in a single bed. On the other hand, a single variety often grew at very heterogeneous stations, just as a single cultivated race could take form under very various external conditions. One and the same physical station did not necessarily favor or produce the same kinds of variations in a single species, and contrasting stations did not always find a parallel in the kinds of variation.[15]

Botanists argued that these two sets of facts indicated the presence of other, external causes, and continued to state that varieties were brought forth by peculiarities of climate and soil. On this basis, though, they could not explain why some variations were ephemeral while others became more or less permanent. To explain how accidental aberrations caused by extrinsic effects could become more or less permanent, botanists would have had to admit inner causes, and this, as Caspar Friedrich Wolff pointed out, would have meant that the essential character of the species determined the formation of the variety.[16] Not only would this have removed any difference in principle between specific and varietal or between constant and variable characters, it would also have established a causal relation between species and variety.

Parallel arguments were constructed around the phenomena of hybridization. So-called bastards were classified as a special kind of anomaly resulting from crosses between individuals with different organizations.

"When a female of one species is fertilized by the male of another species, bastards appear whose formation is constituted from that of both parents alike."[17] This placed bastards outside the rules ordinarily followed in reproduction. Hence their anomalous position and the strict limits imposed upon bastardization. Each separate species had been designed to function independently yet contribute to the overall economy of nature. Intermediate forms lacked this teleological justification, and naturalists reckoned among the wiser provisions of nature the rarity of crosses between species in their free, natural state and the fact that bastards were infertile. Exceptions to these rules were either rare or artificial. Crosses between parents with very different structures were flatly impossible, and Linnaeus's conjectures about crosses between orders were idle speculation.[18]

Obviously, these limitations did not apply to crosses between varieties of a single species, since they often produced new races, even in the state of nature. Again assuming the fixity of species, botanists were primarily concerned to establish whether specimens were crosses between species or between varieties. The possibility that with regard to sexual affinity there might be no great difference between species and varieties was ignored. Observations and experiments were read not on the basis of their inner coherence but according to the traditional principle, in order to arrive at the foregone conclusion—that species hybrids were absolutely or relatively sterile.

The obvious point of entry into the study of plant hybrids is the competition on the question of plant sexuality proposed by the St. Petersburg academy in 1759.[19] The competition summed up received opinion on plant hybrids and initiated research that marked the beginning of serious inquiry into the hereditary characters of plants.

By the mideighteenth century, the question of whether plants have sexual organs and reproduce sexually would seem to have been decided, but as Delaporte has shown, this was not so.[20] Many botanists still contested the idea. To end the conflict, the St. Petersburg academy proposed its question and received three answers; two were judged unsatisfactory, the third was "praemio omnino dignus." At their September 6, 1760, session, the academy awarded the prize to Linnaeus.[21]

Linnaeus's essay contains nothing new; he simply restated positions he had defended throughout his career: in plants there is no fruit without flower; since the flower is a necessary antecedent to fruit, sexual organs must exist within the flower; and those parts with the rudiments of fruit, on the analogy with the animal kingdom, must be female (ibid., pp. 113–

18). Linnaeus then separated and combined male and female parts to show that seed developed only when pollen contacted the stigma. Finally, Linnaeus argued that the fertilization of a female plant by a male of another species exhibited very clearly the sexual duality of plants (ibid., pp. 119–26).

Linnaeus considered the production of hybrids a decisive proof of sexuality, but he offered no adequate proof that the hybrids he cited were offspring of separate species. One, *Tragopogon hybridum*, had been hand pollinated and marked with a thread. Three other plants, *Veronica spuria*, *Delphinium hybridum*, and *Hieracium hybridum*, had been found in the wild or in gardens and exhibited characters intermediate between two known species. Linnaeus assumed these plants would reproduce through seed to form constant varieties. "It cannot be doubted," he wrote, "that we have here a new species, brought forth through hybridization" (ibid., p. 127). The bastard plant, resembling the father outwardly, was with respect to inner medullar substance and fructification the image of the mother. It followed, he argued, that the species belonging to one genus were in the beginning a single species and that the other species appeared through hybridization. In other words, a genus was the epitome of plant forms stemming from a single mother and various fathers.

However plausible Linnaeus's hypothesis, the facts were that crosses between species were difficult to produce and impossible to perpetuate, either because of their absolute or relative infertility or because of the regressive degeneration to which their issue was subject, requiring for fertilization the aid of one of the parental stocks. The barriers of sterility, degeneration, and distribution limited the effects of crossing far more than Linnaeus allowed, and for each species known to be distinct and constant, most naturalists continued to assume a common origin and epoch. The key word is, of course, "assume." The constancy of specific characters and the absolute or relative sterility of specific crosses were subjects about which naturalists found themselves obliged to make assumptions. They were at a loss to ascertain characters that breeding experiments alone would have revealed. They classified bastards as anomalies, outside the rules ordinarily followed in reproduction.[22]

An adjunct of the St. Petersburg academy, Joseph Gottlieb Koelreuter, had never doubted the possibility of these unnatural productions. Unlike Linnaeus, Koelreuter believed that nature restricted their production, thus preserving the order that had reigned in Eden.[23] Two different animals in a state of nature did not produce bastards; nature avoided such disorder by means of instinct. She had equally certain methods to avoid a corresponding disorder in plants—distribution, for example. Bastards were products of those unnatural and artificial worlds of man's creation, botanical and zoological gardens. Even in these unnatural conditions, however, Koelreuter was convinced that nature limited disorder, and he set to work to discover those limits.

Koelreuter produced his first plant bastards in the fall of 1760. Offspring of a cross between two tobacco species, *Nicotiana paniculata* and *N. rustica*, flowered the following spring, and in the fall of 1761, Koelreuter produced an account entitled *Vorläufige Nachricht von einigen das Geschlecht der Pflanzen betreffenden Versuchen und Beobachtungen*. He reported the results of further experiments in three *Fortsetzungen* (1763, 1764, and 1766). The *Vorläufige Nachricht* and the *Fortsetzungen* offer an account not only of experiments in hybridization but of the processes of pollination and fertilization. Koelreuter also published separate papers on these subjects in the *Commentarii* of the St. Petersburg academy.

The *Vorläufige Nachricht* continued the discussion begun by the prize competition. Koelreuter's format, an analysis of fructification followed by experiments, is the order prescribed in the academy's question. Again, the production of plant hybrids constituted the decisive argument for sexual duality. On this point, but only on this point, Koelreuter's argument paralleled that of Linnaeus. Koelreuter had otherwise taken pains to distance himself from Linnaeus's wild claims and speculative flights.

Koelreuter began by rejecting ovist and spermist versions of preformation. He rejected Linnaeus's theory of medullar and cortical layers as well, which he thought more clever than correct. The organized body was, rather, a product of two fluids determined by the Creator for union with one another. The union of these fluids resulted in a compound fluid possessing a compound force equally determined by male and female agents. This compound force, a purposive agent, was the source of organization. For each class of organized beings, a specific compound force produced a determinate structure and a specific nature (ibid., p. 42).

Nature worked in the same way in a cross between species. Once the male and female agents united, formation proceeded rapidly from the nucleus to the flower. The resulting bastard had, however, a twofold nature, being composed from the seed matter of two separate species, and this twofold nature revealed itself in the bastard's intermediate form and its absolute or relative sterility.

When Koelreuter produced a cross between *Nicotiana paniculata* and *N. rustica*, he experimented with many flowers at different times. Each time, fertilization resulted in perfect, although somewhat divergent, seeds. From 110 seeds he grew seventy-eight plants; he kept twenty-one plants over the winter, and in March 1761 they flowered. The plants revealed their hybrid nature in the "geometrical proportion" between differences in the parents, measured in the spread of their branches, the situation and

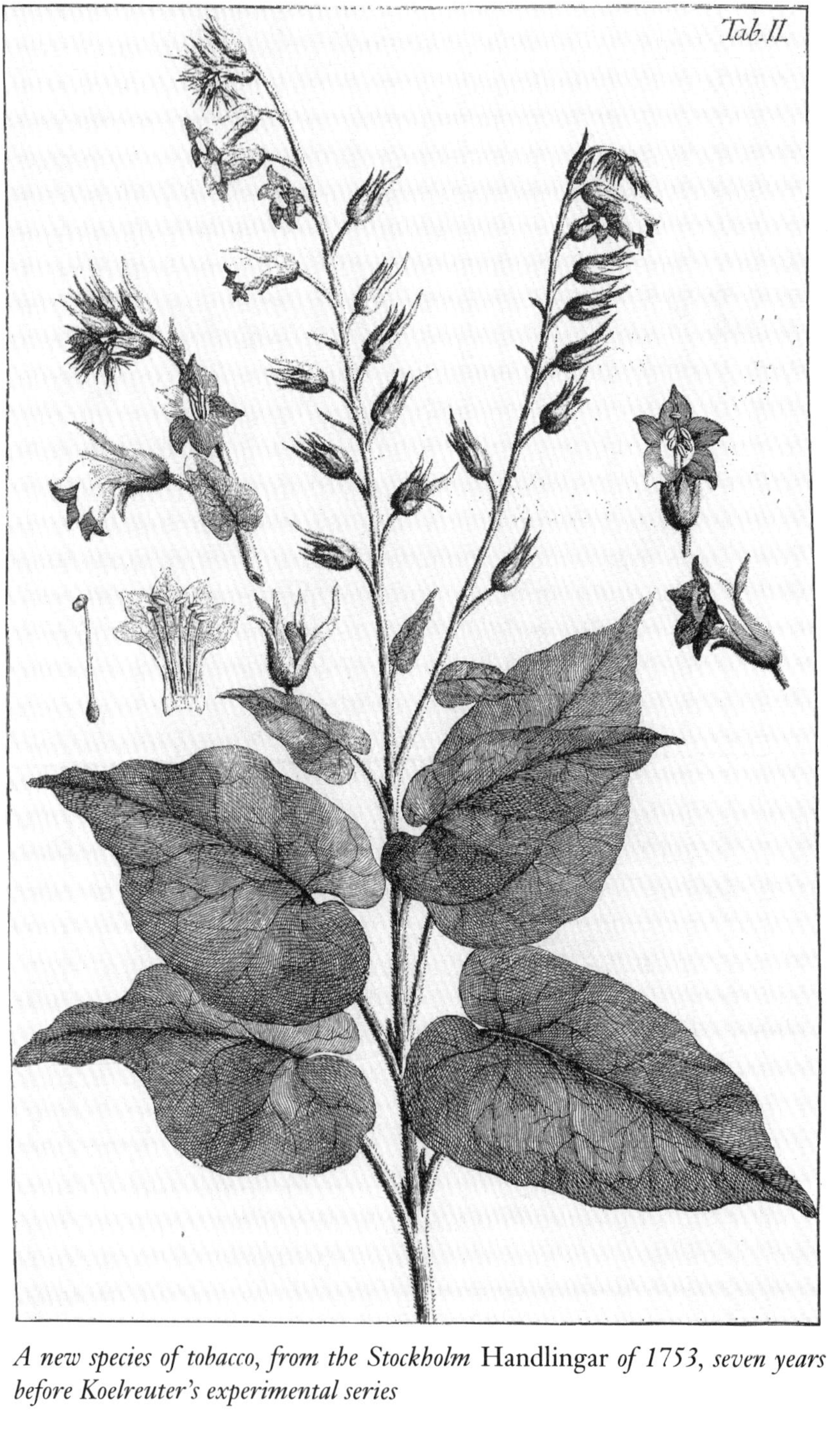

A new species of tobacco, from the Stockholm Handlingar *of 1753, seven years before Koelreuter's experimental series*

color of the flower, and in the individual floral parts. Experiments with seven other genera supported the inference of intermediate size in first generation hybrids (ibid., pp. 69–70).

All offspring of the cross between *N. paniculata* and *N. rustica* were identical. When Koelreuter reversed the direction of the cross, *N. rustica* × *N. paniculata*, the offspring "agreed with the plants of the first experiment and reacted in the same way to the experiments performed on them" (ibid., pp. 53–54). This phenomenon, now called the identity of reciprocal crosses, refuted one belief concerning parental contributions to offspring. The hybrids with which naturalists were familiar showed a greater resemblance to the mother than to the father, and many naturalists, but not all, assumed this was a universal character of hybrids.[24] Koelreuter's reciprocal crosses and detailed measurements established that parental contributions were not so easily distinguished. Robert Olby has pointed out that Koelreuter did not come across any example of sex-linked characters in his experimental plants.[25]

A second essential character of bastard offspring was their absolute or relative infertility. The pollen containers of the *Nicotiana* bastard were small and shrunken and held scarcely any fluid. Not one of the many flowers bore a single seed, even when dusted with a large quantity of the plant's own pollen.[26]

Infertility was one of those imperfections, however, that, when regarded from the vantage of ultimate consequences, proved to be a positive good. By means of sterility, nature preserved the order established at the Creation. "What an astonishing confusion would the peculiar and unchanged hybrid characters and the consistent fertility of such plants give rise to in nature? What a monstrous swarm of imperfections would they bear, and what evil and inevitable consequences would ensue?" wrote Koelreuter (ibid., p. 44).

Koelreuter's evidence for infertility and progressive reversion parallels work on animals published by Buffon three years later. Under certain conditions, a male mule could engender progeny and a female mule give birth. Buffon considered the productivity of different animal species and found a wide range. Fertility varied, and just as species differed in productivity, so too must hybrids, a conjecture that harmonized with Buffon's transformist position. But Koelreuter regarded the species as products of the Creator's hand, and when his experiments produced hybrids with varying degrees of fertility, he read the results as potential reversions to the original species (ibid., p. 76).

At an early stage in his experiments, Koelreuter found a striking contrast in the uniformity and stability between first- and second-generation hybrids. First-generation hybrids for any single cross were all alike: intermediate in form between the parent species and wholly or relatively sterile. Second-generation bastards, even when taken from a single ovary, were all different; they tended less toward the bastard parent, more toward the grandparent (ibid., p. 72). Crosses in which the two seed matters were in equal proportion produced second-generation bastards resembling the first-generation parent. Crosses of seed matters in unequal proportions produced second-generation bastards resembling one or the other original natural species, depending upon which seed matter dominated and to what degree. In other words, Koelreuter indicated roughly the three segregating classes for second-generation hybrids (ibid., p. 30).

The contrast between first- and second-generation hybrids required only a simple adaptation of the theory of normal reproduction. In natural fertilization, the two seed matters and the simple forces in them united to form in equal proportions an intermediate product with a corresponding compound force, hypostatized during the process of generation in the specific characters of external form (ibid., pp. 42–43). Among all original species there were degrees of affinity. In species where affinity was remote, where characters and forces were very different, crosses did not take place. Where affinity was close, a sexual affinity also existed, and crosses might take place (ibid., p. 78). In the first generation, the unnatural state of the plant bastard revealed itself in an enhanced vegetation and curtailed reproduction. Where infertility was not absolute, the unnatural state of affairs revealed itself in the wide variability of the next generation. The principle of equality was shattered, and the seed matters combined in different proportions to produce "all kinds of wrong paths" (ibid.).

Here, too, Koelreuter's speculation paralleled informed contemporary opinion. In a discussion of domesticated animals, Peter Simon Pallas noted that inconstancy of form, once introduced, grew from one generation to the next. Pallas supposed it to result from a vice introduced into the generative faculties of the species by way of crosses.[27]

The presence of essential characters resembling the original natural species was Koelreuter's key to the contrast between first- and second-generation hybrids. He concluded that, although new varieties might arise through selfing, plant bastards did not establish the new and constant species Linnaeus had predicted.[28] Bastard races must revert to the parental species, and Koelreuter undertook to reproduce the situation experimentally.

Offspring of *N. rustica* × *N. paniculata* were intermediate in form, produced sterile pollen, and were only somewhat fertile on the female side. Koelreuter pollinated these with *N. paniculata*. The plants grown from the seed resembled the pollen parent and grew distant from the maternal form. The next summer, Koelreuter pollinated the second generation with *N. paniculata*, and in 1763 sowed 128 seeds, most of which germinated.

When they bloomed, they so resembled *N. paniculata* "that if one had not marked them with particular numbers, one would have differentiated them with difficulty."[29] Fertility of pollen and seed had increased markedly, although this fluctuated. *N. paniculata* was well on the way toward gaining the upper hand with *N. rustica*, and Koelreuter foresaw that this series of experiments would end by producing true *N. paniculata*. "In a word," he wrote, "I do not doubt the possibility of transforming one species into another" (ibid., p. 135).

In the summer of 1765, four years after he had initiated this experimental series, Koelreuter produced the first complete transformation of *N. rustica* into *N. paniculata* in the fifth generation or, in his own words, "in the fourth ascending degree." It was, he wrote, as if "he saw a cat appear in the guise of a lion" (ibid., p. 66). The opposite process—reversion in the direction of the original mother plant, which he followed in a parallel series of experiments—would, he predicted, require a different but proportional number of generations. He also announced the transformation of *Dianthus* and *Mirabilis* species. In all cases, Koelreuter produced offspring from a cross *AB* indistinguishable from the natural species *B*. But since the maternal plant of *AB* was *A*, he felt he had transformed species *A* into species *B* (ibid., 199).

The analogy for this experimental series was the transmutation of metals. In older doctrines of substance, the "real" properties of a body were hypostatizations of striking sensuous qualities. In theory, one body could be transformed into another by taking from it particular properties, conceived as so many independent substances, and substituting other properties for them. Mercury could be ennobled by removing the two characters upon which its fluidity and volatility rested and substituting other characters for them. In the process two seed matters were required: the male seed matter, which was of a sulphuric nature, possessed the force to make the fluid, mercurial female seed matter capable of resisting fire and of forming a stable body. The result was a metamorphosis, in which the male sulphuric agent asserted its superiority.

In the millenia during which transformations of this kind had been attempted, it had occurred to no one that one organic body might be transformed into another. Koelreuter believed he had carried out just such a transformation in the course of a few years. In plants, the male seed matter was an oily sulphuric substance; union with female seed matter produced a stable, organic body, "the first basis of the future plant." In succeeding generations, the male agent got the upper hand, and in the end, the nature of the female had been transformed. The transformation depended upon "close affinity, fertility not wholly suppressed in the production of bastards, and dominance in a certain degree" (ibid., p. 88). Alchemists might have reached their goal, Koelreuter argued, if they had set up an experimental series and observed a few rules with regard to the affinity of seed matters. It was also probable that one species of animal could be transformed—a canary into a linnet, for example: experiment had already established that the female retained fertility in "the second descending degree," and it was therefore probable that she would do so in the "second ascending degree" (ibid., p. 89).

The change of characters in the product of any unnatural union reflected the proportions of the two seed matters. Variability increased in the offspring of bastards, since the possible combinations of the two seed matters were infinite, and nature lacked that thread of Ariadne, natural affinity. Koelreuter did see limits to variation, however. By means of crosses, a naturalist could transform one affine species into another or produce one of the infinite number of possible variations between the forms of the two species. But he could not hope to produce wholly new species with entirely new characters. The naturalist transformed natural bodies by altering or removing particular characters or by substituting other characters for them.

Koelreuter read his experiments over against a conceptual tradition that did not allow for development away from the original natural forms. Moreover, he considered plant hybrids to be artificial products, able to survive only in artificial conditions. Even if one found a true bastard in an open field, one would have to question whether there had not been some human interference. This was an argument Koelreuter constructed around *Verbascum*, where the likelihood of natural crosses was very great (ibid., pp. 192–93). He did not believe, however, that unnatural crosses could transform natural species, even when the natural species crossed regularly. In an open field, "the question would remain whether this chance had not taken place in a region where the natural situation as a whole had been changed or destroyed: true wilderness as it comes from the hand of nature is one thing; a field, free but in respect to a hundred things often very much altered by the hand of man, is another" (ibid.). In an open field, a plant would accept the male agent intended for it, excluding alien matter from fertilization. Nor was there any mention in the writings of ancients or moderns of any bastard *Verbascum* having arisen in the field. Linnaeus had mentioned a *Verbascum* bastard, but this was probably a product of the unnatural conditions of the Uppsala botanical garden. In any case Linnaeus's plant had not produced seed (ibid., p. 189).

The intermediate character of plant bastards, the identity of reciprocal crosses, and the transformation of one species into another, exhibited tellingly the nature of biparental heredity. However artificial the conditions under which species crosses took place, the very possibility of such crosses refuted all theories of monoparental heredity. The two sexes contributed equally to their offspring. Neither the male nor the female agent acted exclusively, although one might get the upper hand, and the two agents always and necessarily acted together.

The functional continuity of Koelreuter's work consisted not in a concern for the laws of heredity but in a stubborn defense of the traditional essentialist species concept. He set out to prove that a cross between two species does not produce new and constant species. Moreover, he regarded the essence of the species as monolithic. "At no time," Ernst Mayr has written, "did he partition the phenotype into individual characters and trace the fate of a given character in different combinations through several generations."[30] Second-generation and back-crossed hybrids did not lend themselves to the same neat generalizations. Here the crucial element was the presence of forms resembling the original natural species. From this fact, Koelreuter argued that hybrid offspring would sooner or later revert to one or the other of the original species.

Contemporaries were only too happy to accept Koelreuter's conclusions. The rigor and complexity of his experiments were alien to the mentality of many naturalists of the later eighteenth century, but his results supported very powerfully the theory of limited variability then being elaborated. Nor was Koelreuter's work so obscure or inaccessible that no one read it. As the *Vorläufige Nachricht* and three *Fortsetzungen* were published, a reviewer in the *Göttingsche Anzeigen* astutely picked out the tendency of the argument: not only had Koelreuter limited the possibility of novelty by means of species crosses, he had shown that change was a narrowly defined combinatory of already existing characters.[31]

These summaries and memoires published by Koelreuter between 1770 and 1796 in the *Commentarii* and *Acta* of St. Petersburg were read, assimilated, and commented upon by two of the most influential and widely read naturalists of the period, Johann Friedrich Blumenbach and Peter Simon Pallas. In an early important work on the theory of generation, Blumenbach presented Koelreuter's experiments as a refutation of preformation. The possibility of specific bastards contradicted any theory of preformed germs, while Koelreuter's transformation of one species into another "must undeceive even the most partisan defender of the theory of evolution [i.e., preformationist] from his prejudice."[32]

Pallas claimed that Koelreuter had refuted the opinions of Linnaeus. He continued:

> The great difficulty we have in producing crosses, with all our human industry, between two different species, the impossibility, confirmed by fact, of perpetuating these as crosses or distinct races, either because of their

> absolute or relative infertility, or because of the regressive degeneration to which their issue is subject, since these crosses require for fertilization the aid of one of the primitive branches, all this opposes facts to a simple probability and forces us to see that all of those species that nature affects to render alike as primitive [were] projected in the first plan of creation, and destined to form that chain of beings that we admire without being able to account for it, any more than [we can account for] the choice, harmony, and combination of colors and ornaments that the same creative force has used to establish its works.[33]

After the completion of Buffon's *Histoire naturelle* in 1767 and the tenth and twelfth editions of Linnaeus's *Systema naturae* (1758–59 and 1766–68), several German professors undertook new dispositions of the mammals.[34] The professors were inspired by the possibility of a rigorous formal classification, on the basis of either essential characters or total habit. Their methods worked against Buffon's arguments for reproductive criteria in determining species and for descent in interpreting the higher categories. Only bits and pieces of Buffon's apparatus—his eloquent descriptions and anatomical measurements—entered general descriptive practice. With the exception of degeneration, Buffon's genealogical terminology was ignored or was implemented only in the limited case of domestic animals. Buffon's argument for the physical existence of species in the succession of individuals who reproduced among themselves was given the usual interpretation: that species characters were constant and that species members bred true. No one at work on the disposition of mammals, with the exception of Pallas, grasped the idea—that this chain of successive existences was the basis for what Buffon called *physical truth*, in contrast to the arbitrary abstract truth based upon affinities and analogies.

The reasons for so massive a rejection are not hard to find. The mammalian table of organization—Erxleben's *Systema*, for example—was pieced together from uniform structural analyses of the mammals then known. Each analysis, limited to the dimensions of number, figure, proportion, and situation, set before the reader a purely visual representation of an ideal type. Buffon's practice in this area did not alter the procedure. He too believed that the method of inspection must bear upon the form, the size, the number, and the position of parts.[35] The naturalist grouped these types into genera and orders on the basis of purely external structural likenesses and differences, intending, if not attaining, a natural disposition of the entire group. Works of this kind, an identifiable scientific genre concerned solely with structural types and their affinities and analogies, necessarily treated information about descent and geographical distribution as circumstantial and secondary and relegated them to lists of variations after the true species or to brief notes on habitat.

Buffon's species criterion, fertile interbreeding, was treated as an alternative to morphological criteria—a particularly inconvenient alternative, at that, something mentioned in prefaces and introductions and promptly discarded. Buffon's suggestion that neighboring groups in the table of organization too distant to allow crossing might have descended from a single branch, could be regarded as one possible interpretation of the word *affinity*. What could be established without argument, of course, was the presence of external likenesses and differences, however interpreted. Only as the procedures of comparative anatomy were gradually adopted, and after zootomists had repeatedly demonstrated concordances of plan in animals, did naturalists turn to the idea of a prototype, "a first design," as Buffon had stated it, "upon which everything seems to have been conceived."

In this situation, only Buffon's actual systematic dispositions, the weakest link in his arguments, seemed relevant to the needs of working naturalists. Every specialist could list Buffon's blunders. Buffon had reduced domestic sheep to the same stock as the *Bouc-étain*, the *Chamois*, and the *Brebis* with only one viable bit of evidence, the production of fertile offspring from a cross between the *Bouc* and the *Brebis*. Buffon had taken the genus of gazelles for variations of one stock, although he knew that gazelle species did not mix in a state of nature and were as different among themselves as the carnivora with retractile claws so scrupulously distinguished in the *Histoire naturelle*. Buffon had, for example, proposed the American anteater as the origin of the *Pangoulin* and the *Phatagin*.

A memoire on variation in animals, delivered by Pallas at the public assembly of the St. Petersburg academy in 1780, offers an analysis of Buffon's ideas that allows us to penetrate the thought behind the more or less complete rejection of Buffon's revision of the mammalian table of organization.[36] Pallas conceded that Buffon's genealogical method had rendered a service to zoology with respect to domestic animals, but Buffon had gone on to propose the transformation of certain natural species into neighboring affine species, and he had added even stranger views on the origin of the mammals in the New World (ibid., pp. 77–78).

Pallas argued that the three principal causes upon which Buffon had rested his hypothesis of degeneration—climate, nourishment, and accidents multiplied by the number of individuals—had never changed the form, proportion, or inner structure of any species to a degree that would sustain his hypothesis (ibid., p. 78). Animals transported to contrasting climates changed only the quality and color of their hair. Nourishment changed only the cut and proportion of horns and the layers of fat. Under altered conditions, these changes disappeared in a few generations. And in the species of quadrupeds and birds there existed aberrations that could not be dependent on the number of individuals in the species. Some very numerous species did not vary, while others, less numerous, varied greatly. Closely related species like the *Bobak* and the *Souslik*, for example, were both distributed in different climates, the herbivorous *Bobak* did not vary, while the carnivorous *Souslik* varied greatly, contrary to Buffon's prediction. Even if one supposed the transplantation of species to be very old and the influence of climate to be greater than it is, one still could not explain why so many species occurring in both the Old and the New Worlds and spread from the tropics to the poles had not varied and did not change their natures continuously. Humans, who were naturalized in every climate, had experienced only superficial changes of skin and hair—and this in spite of their total abandonment of the state of nature with respect to food and habit.

Domestication and its attendant conditions, Pallas continued, were more powerful causes of variation than anything experienced by a wild animal, and yet humankind had not been able to denature any domestic animals to form distinct and constant races or species. Domestic animals whose species were unique in the climate where they had been tamed—the horse, the cow, the camel, and the dromedary—and whose progeniture remained pure experienced almost no changes in the climates in which they were forced to live and reproduce (ibid., p. 82). If, then, the dog, the sheep, the goat, and domestic fowl had undergone marked alterations, these must be due to a far more powerful cause than domestication or climate. "They are the effect . . . of a vice introduced in the generative faculties of these races by mixing species. Vice that could be compared in some fashion to the alteration of fluids and even the action of solids in a living body through the ferment of some miasma" (pp. 82–83).

Just as Koelreuter had restricted the effects of plant hybridization, Pallas narrowed the possibilities for crosses between animal species in the state of nature. The original species did not mix. Instinct kept them apart. When two individuals of different species and sex were enough alike to make an alliance possible, their offspring were sterile or unable to maintain the ambiguous form by themselves. By crossing with one of the parental lines, however, they resumed the original form of the parent in one or two generations. In Russia, for example, the Chinese goose (*Anser Cygnoides*) often mated with the domestic goose. The result was a fertile race that returned to the state of the domestic goose if not crossed anew with the Chinese goose (ibid., 97–97, 83n). The production of new species by means of spontaneous crosses was improbable even among insects and fishes, where heterogeneous fertilization had been observed (p. 84n).

The conditions to join two different species in nature must always have been rare.

Pallas framed his case for domestic animals very differently. He turned to the age when humans first subjugated animals. The survivors of this process had mated with males of neighboring species offered by chance. Offspring of impaired fertility could not transmit the vitiated form, but when crossed with parental lines, they again approached the original forms. Thus the horse, ass, cow, pig, camel, dromedary, reindeer, and cat had remained the same in every climate and had not degenerated in the domestic state (ibid., pp. 82, 84–85). Goats, sheep, dogs, and domestic fowl, on the other hand, had so degenerated under the human hand that it was difficult to recognize the original forms. Some seemed so wholly factitious that no single origin could be ascribed.

The best method for understanding these phenomena, thought Pallas, was that of Buffon, who had established "with the most probability the origin and cause of the numerous and great varieties among these domestic animals" (ibid., p. 86). For each of these four examples, Pallas indicated a principal stock or stocks and a probable mate with which the principal type produced offspring. These specific crosses, often repeated, and having become consistent under the human hand, "introduced the first inconstancy, the first germ of deviation from the primitive type" (pp. 88–89). The resulting races, transported from their original habitats or unable to mix with wild stock, gradually mixed with one another. Inconstancy of form, augmented by external conditions, increased from generation to generation. Finally, these domestic animals became what they are now, variable and bizarre races.

The most complex of Pallas's arguments concerned dogs. Pallas could draw upon a large body of eighteenth-century research to prove that dogs mated with the wolf, the fox, and the jackal and produced fertile offspring. Buffon had counted thirty races of dogs, seventeen of which he had mistakenly ascribed to climate. Pallas advocated instead "adopting as primitive races those which owe the most to wild animals, even those for which modern experience has not yet verified the crossing and fertility with our domestic dog" (ibid., p. 93).

A dog belonging to the nomads of Asia was the first of these races, the *Canis Moloticus* of Aristotle. The jackal may have first crossed with the wolf in the temperate climates of Asia. Earlier, in the north, where the jackal could not survive, couplings with wolves transformed a domestic race of dogs and produced the *Canis pecuaris* of Aristotle. The same dog crossed with the jackal in central Asia (ibid., p. 94). Another primitive race, the bulldog, originally of India and Asia Minor, was probably a product of the hyena and a powerful race of dogs. The offspring resembled the hyena in many ways and, with one or two exceptions, were very unlike other dogs. Education and artifice had since produced the Dalmatian and the hound; a hereditary malady, joined to these causes, resulted in the basset hound. The *Chien Alopécide* common in northern countries was undoubtedly a primitive race, a product of crosses with the fox; through degeneration and the influence of climate and nurture, the mixture had produced the many races of small dogs. Finally, an Icelandic dog descended from the *Isati*, the American dog, from the species of the American fox. These great primitive races of dogs were all products of a mix with a pure domestic species. Many secondary races had been formed from these primitive races by the choice and crossing of individuals, by hereditary faults, by climate, and by nurture (p. 96).

Pallas constructed parallel arguments for domestic sheep, goats, and fowl. In all cases, crosses of affine species, one wild and one domestic, produced fertile offspring and altered the purity of origin upon which the constancy of form depended. Once the generative force had lost its original simplicity, it was liable to produce, under the human hand and the influence of climate and nourishment, greater and more sudden deviations from the primitive type, deviations never found in any pure race (ibid., p. 90). A contrasting phenomenon, the ennoblement of species through choice of breeding stock, did not alter the argument. In many cases, ennobling a domestic strain was only the bastardization of natural qualities. In other cases, most notably the horse, the species was larger and more vigorous than in the wild because breeders took greater care to maintain the strain; without this care, individuals would again approach the natural state (pp. 98–99). The evidence, Pallas concluded, pointed to the fact that

> *the influence of generative forces* surpasses and balances all the other causes that can change the exterior of animals . . . but that these same generative forces, subsisting above all in all their purity, the type formed by the female, not being altered by alien vital ferment, maintain in wild animals and in some domestic species the primitive forms in opposition to all the influences of climate and of nourishment and oppose themselves to degenerations and degradations of the species. (ibid., p. 99)

When a deviation became hereditary, nature reestablished the original form if humans did not intervene by choosing couples and affirming the deviation. This did not occur in nature, because deviations were too rare to meet. And in the natural state, the instinct of females caused them to prefer the most perfect and masculine males, thus preserving the external form of the species.

Left to herself, nature prevented the adulteration of the generative faculty with the insurmountable barriers of instinct, sterility, and distribution. Spontaneous races either did not cross or were not permanent. Human art had obtained fertile crosses, but these products could not pretend to the validity of the parent species "when these two species propagate quite distinctly in the wild state through individuals of two sexes and when these two sexes, joined, differ essentially by some constant characters from the type of the neighboring species" (ibid., p. 101).

Pallas concluded his memoire by contrasting, on the one hand, the factitious variability of domestic species, where the cause of variation was an inner agent, a vice in the generative faculties, and on the other hand, natural variations, where changes in color, cut, and proportion—all nonessential changes—were a consequence of such external agents as climate, nourishment, and disease. By linking natural variation to the assumption that changes could result only from external influences, Pallas placed the constancy of natural species beyond question, After comparing Buffon's hypothesis of degeneration with the facts presented by nature, Pallas concluded that "we must renounce this supposed degeneration of the species, which no other influence has been powerful enough to produce, and we must admit, for all of the species that we know to be sufficiently distinct and constant, a common origin and epoch" (ibid., pp. 100–101). This, claimed Pallas, was a far more plausible explanation "of the general causes of variation among organized beings, of the reason for the remarkable inconstancy of certain races of domestic animals, and of the originality and invariable preservation of what were called species" (p. 72).

These conclusions agreed with those of another work published at this time, E. A. W. Zimmermann's *Geographische Geschichte*. Zimmermann interested himself in the relation between external physical factors and the distribution of animals. He conceded that the existence of variations caused by external physical conditions was beyond question and that some of these variations had become hereditary. But this did not mean, as Buffon had claimed, that physical conditions could produce new species.[37] Zimmermann argued, in direct opposition to Buffon, that every kind had a place and had been "distributed over the earth from the very beginning, each situated in a climate appropriate to itself" (3 : 192; see also 1 : 23–24, 25). Zimmermann conceded that some original kinds had been lost in the wake of catastrophes and that wholly new kinds had formed, he also admitted the possibility that in similar situations some species had approached others so closely that their crosses brought forth new kinds. The present distribution of mammals, then, beyond God's initial distribution, was determined by physical factors and the history of the earth (1 : 26–28).

These truths could not be used to justify Buffon's account of endemism in the New World, that is, the degeneration of Old World forms in the enfeebling conditions of the New World. The degeneration of animal species through climate offered far more difficulties than Buffon appeared to think. Nor was Buffon justified in concluding that wholly new species had appeared. This, wrote Zimmermann, was an entirely new question and one not answerable in the lifetime of a single observer (ibid., 1:25).

On the topic of the general causes of variation, the best-informed naturalists found that Buffon had not made an adequate case for the concept of degeneration. The three principal external causes cited by Buffon for degeneration did not in observable cases change the form, proportion, or inner structure of an animal to a degree that would support the hypothesis of degeneration. This left a single inner cause—reproduction—as a basis for degeneration. Here, too, the observable effects were far less decisive than Buffon allowed. Degeneration operated only in some species under the human hand—and not even in all of these. Buffon's use of reproduction was useful or necessary for determining origins in these few cases, but true species did not mix in a state of nature. Instinct and distribution kept the species pure; occasional crosses between species were sterile or soon reverted to the parental form. All crosses between representatives of the higher classes were impossible.

Buffon's conjectures about geographical distribution and degeneration in response to outer conditions did not account for the present distribution of mammals. The alternative hypothesis of an original distribution, each kind situated in a climate appropriate to itself, not only offered a more plausible explanation of the actual state of affairs but harmonized as well with the then-dominant belief in the constancy of species.

These were serious objections to Buffon's arguments for degeneration. In the eyes of eighteenth-century naturalists, however, what was fatal to Buffon's case was his willingness to ignore or manipulate data to bolster weak generalizations. Naturalists recognized, as the greater reading public did not, that Buffon's statements were, in the words of Georges Cuvier, "based on facts susceptible of contestation."[38] Here is one example cited by Pallas: the American peccary, with the external characters of the pig, had in addition a stomach of several pouches, as in ruminants, and an open follicle on the kidneys, the only other example of which occurred in an African gazelle. Buffon, said Pallas, "treats these two great particularities, which are not the only distinctive marks of the peccary, as secondary characters and believes he can deduce his race of Cochins from the old continent, changed by the climate of the New World."[39] For men trained as systematists, this was an example—and there were many such—of a tendency to agglutinate particulars in support of dubious general ideas.

Although the human species had long been a part of the mammalian table of organization, anthropology remained a study apart, one that only occasionally touched on other mammals. When Pallas began publishing systematic works in the 1760s, the parallel study of humans included discussions of the possibility that the species actually consisted of separate species of different origin, a question discussed since the time of Paracelsus and Caesalpino. Throughout the eighteenth century, polygeneticism enjoyed a vogue among men like Voltaire and Hume, who were not, however, particularly interested in the scientific aspect of the question. Polygeneticism emerged as a topic for serious discussion among working naturalists in the 1760s. Linnaeus divided the genus *Homo* into two distinct species: *Homo sapiens* included four earlier varieties, and *Homo troglodytes* compounded older, mostly fabulous races.[40]

Later versions of this classification alternately included and removed a range of primates and mythical creatures, including *Homo caudatus* and *Homo lar*.[41] Buffon added to the possibilities with two important memoires published in 1766 in which he equivocated over his breeding criterion for species. "Nomenclature des singes" seemed to hint that fertile crosses between Africans and apes had taken place, while "Dégénération des animaux" accepted the possibility that there might be fertile crosses between original species.[42] Henry Home, Lord Kames, summed up the tendency of the discussion in his *Six Sketches on the History of Man*.[43] Human differences were specific differences because of their constancy and geographical distribution. Buffon's "highly artificial" reproductive definition of the species had been refuted by Buffon himself in his latest memoires.

Immanuel Kant's precritical discussion of race, "Von den verschiedenen Racen der Menschen," grew out of a course in physical geography that he taught many times after 1757. Provoked by Lord Kames's revival of polygeneticism, Kant scheduled a discussion of the human races for the summer half of his course in 1775. A prospectus published two years later opened with a statement of Buffon's reproductive criterion for species: "animals that produce fertile young with one another (however different in form they may be) belong to one and the same species."[44] Kant placed this statement in a context that showed that he accepted Buffon's distinction between a real, physical order of living forms and the abstract order of ideas. "The division of schools proceeds by classes according to likenesses; natural division proceeds by lineages and divides animals by affinities with reference to reproduction. The former produces a scholastic system for memory, the latter a natural system for the understanding: the former intends to bring created beings under labels, the latter under laws" (ibid.).

Since this is an early instance of a German thinker having understood Buffon's conception of physical truth, it is worth inquiring a little more closely into the reasoning behind it. In Kant's introduction to his lectures on physical geography, he discussed three possible divisions of natural forms. A Linnaean naturalist approached nature through the logical division of concepts, dividing mammals, for example, on the basis of locomotion, "a division that I make in my head." A division on the basis of space and time, on the other hand, was a physical division, and it could proceed in two ways. One could describe natural forms solely on the basis of spatial distribution, or one could, in turn, subsume this description under a historical account of nature.

> The history of nature contains the manifold of geography as it has been in different times but not how it is now at the same time, because in that case it would be a description of nature. But if one treats the events of all nature as they have been constituted through all times, we are supplied for the first time with a natural history properly so called.[45]

Kant's prospectus on the human species, his first published account of these ideas, opened with a sharp contrast between description and history. If the natural division of animals rested upon the unity of reproductive force that held for a certain manifold, it followed that the key to a natural method would be the recognition of groups united by descent, not by external likenesses and differences. A logical division, proceeding by classes on the basis of likenesses, created a scholastic system for the memory. A natural division, proceeding in terms of lineages, would create a physical system for the understanding. To give a manifold under law was to give experience an a priori unity and necessity, that is, a basis superior to a mere aggregation classified on the basis of likenesses and differences.

Since humans everywhere could produce fertile young with one another, however great their external differences, it followed, according to the reproductive criterion, that all them belonged to a single species. "For this unity of the natural species," Kant wrote, "which is the unity of the generative force common to them all, we can offer only a single natural cause; namely, that all of them belong to a single lineage, from whence, regardless of experience, all have sprung or might . . . have sprung."[46] Kant proceeded to recast the kinds of human diversity as so many results deriving from the unity of a single reproductive force. A natural species consisted of the hereditary marks harmonizing with its parentage. Kant called the result *Nachartung*. Deviations from the parentage he called *Abartungen*. Where deviations passed a certain limit, Kant called the result *Ausartung*.[47]

Species could be altered by distribution, which determined the outer conditions of life (light, air, soil, and nourishment), or by reproduction, which determined the inner conditions of life. There were then just three paths (a form of argument Kant much favored) by which intraspecific differentia established themselves. A degeneration might be constant in reproduction and in distribution as in the white and black races. Or a degeneration might prove constant in distribution, but not in reproduction; this resulted in *Spielarten*, as in the blondes of Scandinavia and the brunettes of southern Europe. Or a degeneration might be constant in reproduction but not in distribution; this resulted in *Schläge*, as in certain families a closed circle passing along a certain character (ibid., pp. 430–31).

The changes in species that resulted in *Spielarten* and *Schläge* were subject to dispersion, but this was not the case with racial differences. The fact that offspring of a cross between two races were invariably half-breeds (*Mischlinge, Blendlingsart, Mittelschlag*) demonstrated that the parental differentia were racial; that the two races produced fertile offspring showed their common origin and the unity of the reproductive force. Hence the possibility of demonstrating racial differences experimentally.

The four races of mankind, according to Kant, were the white, the black, the Hun, and the Hindoustani. These were so distributed that, in general, race and climate answered one another, a symmetry that made skin color an important expression of deep-seated differentia. Where race and location were no longer symmetrical, Kant explained the disrelation as a result of a migration by a race already become constant (ibid., pp. 432–33). Since Kant conceived races as deviations from one common type and argued that local creations needlessly multiplied causal factors, he assumed that there had been a single original stock. Through migration and adaptation to different climates, this stock had branched out to produce the existing races. The original stock, he supposed, had been one of the mediate forms from which developments could unfold in many directions—and he indicated an origin in the temperate zone in one of the darker shades of the white race (p. 441).

Many naturalists abandoned the problem of race at this point and claimed that racial characters resulted from the effects of environment on individuals and that these characters gradually became hereditary. Kant rejected this solution. The harmony between race and climate, he argued, was a purposive arrangement, a "precaution of nature." The basis of the harmony lay in an innate capacity for adaptation, a matter of germs (*Keime*) and tendencies (*Anlagen*) stored in the creature to meet any contingency. The development of a race was not to be explained by appealing to chance or to universal mechanical laws. External factors were circumstantial, not productive causes for what was necessarily inherited and reproduced (ibid., pp. 434–35). Moreover, the capacity for adaptation was limited; it did not remove an animal from its "original determination" (p. 435).

Providence had determined humans for all climates and soils. In different conditions, they developed differently; for the four races, Kant offered "purposive causes" for the gross changes they had undergone. In a cold, dry climate the body needed blood heat to survive—a quicker pulse, a shorter circulation system, a smaller stature. Moist heat, on the other hand, promoted the vegetative parts of the body; the fleshy parts grew and exhalation increased. Kant then turned to the question of why similar conditions had not everywhere produced similar effects. Racial changes, he argued, triggered by prolonged residence in one climate, elided further transformations through the influence of climate (p. 442).

Kant returned to the problem of race in 1785 with a paper entitled "Bestimmung des Begriffs einer Menschenrasse," in which he restated the distinction between physical descent and descriptive groupings based on formal likenesses as well as his conception of the unity of the human species.[48]

> Animals whose differences from one another are so great that it is necessary to attribute their existence to so many separate creations could, to be sure, belong to the same nominal species, classified according to certain similarities, but never to a real species, in which at least the possibility of derivation from a single pair is required. To determine the latter, however, is the proper business of a history of nature, while the describer of nature can be satisfied with the former. (ibid., p. 98)

A natural type would remain true to its original nature however great the diversity exhibited by its descendents. Nature's original model had been stored with germs and tendencies; under changing conditions, some of these had been activated over time. As a consequence, the original lineage—not a single pair!—had been transformed into many diverse forms, transformations that Kant insisted were purposive. The skin bore the marks of transformation that alone justified division of the natural species into different races. Among all the kinds of diversity, racial characters were inescapably hereditary. When fertile progeny of one race crossed with those of another, the result was invariably a middle kind. Fertility in the offspring indicated the unity of the original parental lineage. To posit different original lineages with hereditary characters gained nothing philosophically and broke up the unity of the natural species (ibid., p. 99).

Kant's reputation in the German-speaking world was approaching its zenith, and one young naturalist thought it worth his while to challenge one of Kant's proposals. Georg Forster could not accept Kant's distinction between natural description and natural history, and he mistook Kant's discussion of lineage. Kant, argued Forster, had misunderstood natural history on many points but nowhere more grievously than in his treatment of the effects of climate on organization.[49] Racial differentia, indeed any organizational process, could be derived from purely physicomechanical causes. E. A. W. Zimmermann had by now established that the surface of the earth was covered with plants and animals from the beginning; organized bodies could not have originated from a single point or a single pair. Humankind, Forster continued, was no exception. Kant himself had admitted that nature had given each racial stock a character and a particular organization in relation to a particular climate. "Doubtless this close relation between a land and its inhabitants can be explained most briefly and easily through the local appearance of the latter. If Africa brought forth her men, and Asia hers, then it is not difficult to conceive why the one and the other match their climate so particularly" (p. 150; see also pp. 397–409).

Two separate stocks, each constituted from autochthonous individuals, had arisen in Africa and Asia. Since the difference between Hindoustanis and whites was not an essential difference, the Hindoustani could be traced to Asia, whites to the Caucasus. America, which had only become inhabitable later, probably had no autochthonous individuals, but on this score, all was uncertain (ibid., p. 151). To derive blacks from a separate stock was not to cut the cord of common humanity: blacks were a brother species, and as in the community of nations, their differences from Europeans served to enhance the potential of both species (p. 153).

Kant's "Über den Gebrauch teleologischer Prinzipien in der Philosophie" answered Forster on three points.[50] The Hindoustani were a separate race because their descendants in Europe, the gypsies, preserved the type over many generations and inevitably produced half-breeds when crossed with Europeans. The cycladic distribution of races showed that racial differences were an effect not merely of purely climatic conditions but of adaptation to external conditions by means of innate tendencies. If like conditions alone elicited racial differences, like races would be found in like climates everywhere. The races, however, were not distributed sporadically but cycladically. Kant's chief reason for answering Forster, though, was Forster's statement that physical causes were sufficient to explain the origin of organic creation. Kant rejected the idea on critical grounds. Forster had abandoned experience and had overstepped the limits of scientific explanation.

All organic beings, Kant explained, derive from other organized beings by means of reproduction. All subsequent differentiation develops from original innate germs and tendencies. The appearance of organic beings remains, of course, inexplicable. Forster, said Kant, had argued that origins would be explained on the basis of factors of an inorganic nature, but his assertion was unprovable and conflicted with the conditions of experience (ibid., p. 179).

Every living creature, said Kant, reveals to observation a material being whose organization forces us to relate every part as both cause and effect to every other part and to posit organizing forces that work purposefully. To the degree that we follow experience, we find ourselves in a dilemma. On the basis of observation, we cannot exclude purposive forces, but neither can we ascribe these to matter or to natural forces. When we posit a force of any kind, it must work either mechanically or purposively. On the basis of mechanical causes, we can never conceive the origin of an organism, nor can we explain it on the basis of a cause acting purposively, since experience of these causes is given only in ourselves. We cannot grasp the appearance of an organism as an object of a determining judgment. Every attempt to determine it is independent of experience. The use of teleological principles must therefore be heuristic, not metaphysical (ibid., pp. 181–82). In his presentation of the human race, Kant believed he had made just such an argument.

In the two later papers on race, Kant altered Buffon's distinction between abstract and physical methods. Kant split the two along the constitutive-regulative line that runs through his mature philosophy. *Naturgeschichte* was not secure knowledge but belonged to the regulative ideas of reason. But Buffon's basic claim, said Kant, that the problems of system were better viewed in terms of history and descent rather than in terms of formal likenesses and differences, remained intact.

> Whether there really is such an affinity in the human species must be decided by observations that reveal the unity of descent. And here it is clearly seen that one must be guided by a determinate principle merely to observe, in order to pay attention to that which might give information of descent, and not merely on character similarity . . . for here we have to do with a problem of the history of nature and not of the description of nature and mere taxonomic naming.
>
> Of the variation of the original lineage there is no more certain sign than the impossibility of producing fertile offspring through the interbreeding of different groups. (ibid., 164–65)

Kant's papers on race clarified Buffon's distinction between abstract and physical approaches to living forms. The justification for a natural classification lay in the derivation of living forms from a common lineage. Systems based upon formal likenesses would be reformed when subsumed under the natural groups determined by common ancestry.

In the early years of his long career at Göttingen, Professor Johann Friedrich Blumenbach was effectively insulated from the influence of Kant's ideas by his orthodox training as a systematist. Early editions of his *Handbuch der Naturgeschichte* defined living forms as collections or series of individuals characterized by an ensemble of distinctive traits, whose transmission was natural, regular, and indefinite. "Natural bodies are always produced by other natural bodies of the same form and kind, so that their existence presupposes an uninterrupted series of like bodies back to the first creation (whom they have to thank for their existence)."[51] Or back at least to the first prototypes, or *Stammältern*. This statement does not differ much from Linnaeus's "Species tot numeramus quot diversae formae in principio sunt creata."[52] Organized bodies in the state of nature exhibited the form they had received from the hand of the Creator, and the formative force preserved "the determinate form and habit of all individual species."[53]

More puzzlingly, Blumenbach suggested that these "original specific structures" were subject to degeneration. *Degeneration* was not an orthodox term but belonged to the alien conceptual world of Buffon. There are other indications that Blumenbach had read Buffon with attention; for example, he cited the breeding criterion used by Buffon to determine species, which he did not feel was adequate. In later editions of *Naturgeschichte*, he stressed the impracticability of applying the criterion, for example, the difficulty of mating Asian and African elephants and the want of reliable information about fertile hybrids. On the other hand, he insisted that if two species were naturally distinct, they would also exhibit different essential characters.

Blumenbach combined the orthodox position with his new theory of generation effortlessly. The *Bildungstrieb*, or formative force, precisely like a preformed germ, determined and maintained the form of an organized body but was itself beyond the reach of scientific investigation. The capacity of the *Bildungstrieb* to respond to external conditions could not pass beyond a certain limit without destroying the economy of the organism. And the possibility of crosses with other species was limited to closely related forms. With this conception of the formative force, Blumenbach believed he could confront any kind of deviation from the natural type and achieve a satisfactory explanation of the observable characters. The well-defined discontinuities that separated one species from another originated in the faculty of generation. No influence outside the faculty was powerful enough to produce constant deviations, and nature hindered the adulteration of the faculty by means of the insurmountable barriers of instinct, sterility, and distribution.

Blumenbach's difficulties came from a somewhat more dubious element in his formative force. His conception of the force united two distinct principles—the mechanical and the teleological. His difficulties with these incompatible principles led him to consider Kant's reformulation of his ideas. As we have seen, Kant used teleological arguments to explain racial characteristics. This in turn was part of a much more elaborate argument for the heuristic application of teleological principles in the limited cases where mechanical principles did not suffice. Tim Lenoir has shown how Blumenbach gradually eliminated terminology that might have led to vitalist interpretations of his formative force, replaced his own account of the formative principles with that of Kant, and took over Kant's definition of the organized body as "one in which every part is reciprocally end and means."[54]

To reconcile Kant's ideas with his own orthodox position, Blumenbach focused, as Phillip Sloan has argued, upon the affinity between Kant's conception of lineage and the formative force.[55] Like a formative force, lineage emerged from the organization of generative fluid and produced, maintained, and transmitted structure. Lineage contained germs and tendencies that allowed for some adjustment to external circumstances, just as the formative force absorbed variations that did not threaten the economy of the organism. The relation between the integrity of the species and their diversity was similar in the two theories. In the end, Blumenbach managed to integrate the atemporal, formal aspects of the lineage concept with his own formative force. He rejected, however, Kant's use of descent and the breeding criterion for the determination of natural groups. The principle of copulation, he wrote in 1795, was not sufficient to distinguish natural species from varieties. By 1797 Blumenbach had accepted the possibility of Kant's breeding criterion, but he added that difficulties of implementation remained.[56]

Prominent among these difficulties was the determination of groups above the species level in terms of descent. Since Kant had argued that only members of the same species produced fertile offspring, advocacy of an entire system based on descent must have seemed contradictory. One possible solution was to draw an analogy between the formation of races and the formation of species. Just as a lineage was a set of germs and tendencies that provided a basis for the different races of man, so prototypes of higher orders may have provided a basis for the emergence of species in response to external conditions. The emergent species would be guided and limited by the germs and tendencies of the prototype. To account for their reproductive isolation, one had to assume that the lineage was altered fundamentally in the process.[57]

When Blumenbach examined relations among species, the evidence for a unity of plan was overwhelming. When Blumenbach studied fossils, he found them different but related to present living forms; changes in external conditions had triggered changes in the formative force that left the type intact and produced a structure similar to the prototype. "By these transformations, organisms in part of similar type to those of the *Vorwelt* are reproduced with forms that are better adapted to the new order of things, since they are produced by the modified laws of the formative force" (cited in ibid., p. 95).[58] Shifts in environment had an effect upon organization and resulted in modifications of the formative force. But the modifications were not to be ascribed to degeneration; they were, rather, the products of the formative force. "Such a fact cannot be a consequence of degeneration; rather it must be attributed to transformation through an altered direction of the formative force" (cited in ibid., p. 96).[59] There were limits to these modifications. The species were not transformed by acquiring entirely new characters; modifications were a consequence of the fact that organized beings tend to vary. "Admittedly, no explanation of this phenomenon can be given. Suffice it to say that the phenomenon itself is the unmistakable consequence of the inherent changeability of nature."[60]

The results of Blumenbach's exploration of the implications of Kant's genealogical method are apparent in the 1797 edition of the *Handbuch der Naturgeschichte*. Blumenbach accepted parts of Kant's program but retained an orthodox framework. Naturalists, he said, would achieve a natural method when investigation of generation and heredity had laid out the laws of form and when geology had revealed the history of the formative force. Blumenbach did not give these ideas a systematic exposition, but he continued to explore their implications. More importantly, he passed along to his students the tension between his own orthodoxy and the reform proposed by Kant and Buffon (cited in Lenoir, p. 96).

Some of his students had already taken a step beyond Blumenbach's cautious practice. Christoph Girtanner of Göttingen echoed Immanuel Kant in the opening pages of his *Ueber das kantische Prinzip für die Naturgeschichte*.

> The history of nature in the philosophical sense divides organized bodies into lineages according to their relationships of generation. It is grounded on the unifying law of procreation. The unity of the species lies in its unity of reproductive force. In this way, a natural system for the understanding is formed, a division of organized bodies under laws, and particularly under the laws of the formative force.[61]

Girtanner's mention of the formative force is only one of his attempts to weld the ideas of Kant and Blumenbach into a single program; on the issue of species criteria Girtanner also followed Kant. Fertile reproduction, he said, not outer likenesses, defined the natural species. "All animals or plants that generate fertile young with one another belong to a physical species. This is the great law on which the history of nature is founded" (ibid., p. 4). Girtanner cited evidence from breeding experiments with mammals, birds, and plants that demonstrated the universality of Kant's criterion. According to Kant, races of one species, however different in outer form, invariably produced fertile young. Backcrosses between these young and the parental stocks extinguished the hybrid form, but otherwise true racial stocks, once established, could perpetuate themselves indefinitely.

Koelreuter's experiments with *Nicotiana* provided a model for these breeding experiments: his production of fertile hybrids "proved" them to be races of one species. Other evidence showed that, after the fifth successive crossing, offspring of Caucasian and mulatto stocks presented none of the negroid characters of their hybrid origin. Girtanner followed the same rationale in classifying dogs and horses.[62]

Applying this racial conception to the species, Girtanner called for a physical history of nature. "Natural history teaches us (or at least attempts to teach us) how the original form of each and every prototype (*Stammgattung*) of animals and plants was constructed and how species have gradually been derived from the prototype."[63] It did not matter so much whether a prototype was a single pair or a group of individuals. The task of natural history was to reconstruct the germs and tendencies present in the prototype and to establish the physical conditions under which some were activated while others remained dormant, thus issuing in different but related species.

Natural history, then, according to Girtanner, grouped organized beings according to lineage. "In this manner a natural system arises for the understanding, a grouping of organisms according to the laws of the formative force" (ibid., pp. 3–4). The breeding experiments and correlations needed to achieve a natural system would "probably take centuries." For this reason, the systems of Linnaeus and Blumenbach were useful but only until, "through exact observations, trials, and experiments, the laws of reproduction are sufficiently known. Then, however, a new division of the animal kingdom must be undertaken according to the relationships of reproduction" (p. 73).

Diversity within the species was rooted in the germs and tendencies of a common lineage. But how were species to be related to a common genus, order, or class? Girtanner emphasized that any relation between species in terms of reproductive capacity involved an alteration in the formative force. Only further research could uncover the conditions leading to variations in the formative force. "When, finally, further advances in the construction of natural history are permitted us and the laws are revealed according to which the organization of bodies is altered . . . only then will we be able to demonstrate how organic nature has come to be in the condition in which we now find it" (ibid., p. 55).

Girtanner's proposal drew a mixed response from working naturalists, many of whom now understood that the principle behind Kant's ideas would reform their science. Although they were prepared to accept a genealogical approach as possible in theory, implementation remained a problem.[64] The reasons for this are clear when we study another case, that of the brilliant young naturalist Johann Karl Illiger. Following Girtanner's lead (Phillip Sloan has shown that Illiger's key terms paraphrase Girtanner, see ibid., pp. 142–43), Illiger distinguished the physical and orthodox determination of species: "We can determine the species only on the basis of experiences about reproduction, and it is an error if we assume, as is usually done, that the species originates from the extraction of common characters shared by several individuals."[65]

Illiger rejected the principle of analogy. When a naturalist found individuals differing from those of known species, "he distinguishes them as separate species. The separation is to be considered provisional, however, and only experience can decide whether the species is founded in nature or not. It would be quite wrong to declare the distinctive character to be the basic criterion of species, since the essence of species consists solely in continuously fertile reproduction" (p. xxix).

Illiger, like Girtanner, declared himself firmly on the side of Kant. This amounted to little more than a programmatic statement, whose effects were limited to the preface of his *Terminologie*. The body of that book was an attempt to relate the physical definition of the species to purely morphological definitions of the higher categories. The shift took place in the preface, which discussed the relation between species and genus. Natural species, defined by fertile interbreeding, it said, also exhibited some general character that revealed the inner and outer relation of parts. On this basis, Illiger shifted observation from the diachronic axis of genealogy to the synchronic axis of morphology as he climbed the systematic hierarchy. He defined *genera*, for example, as the mass or totality of species to which the characters contained in the habit commonly belonged (ibid., pp. xxxiii, xxv, 108). He then applied the principle to his system as a whole: "Just as we form the genera from species, so we obtain the higher genera. The orders and classes are formed from the genera" (p. xlvi).

In the end, Illiger's system did not differ so greatly from that of Blumenbach. Illiger recognized the differences between abstract and genealogical methods, but he could not extend the genealogical reform above the species level.

Men trained in orthodox systematics approached plants and animals as identifiable kinds. They analyzed each kind according to a set pattern, describing each part in turn according to the dimensions of number, figure, proportion, and situation. The result was a static visual representation that set before the reader an ideal type. The types were then grouped in a table of organization that both situated and contained them. The end product, said Goethe, was "a kind of mosaic in which one finished piece was placed next to another in order finally to produce out of a thousand individual pieces the semblance of a picture."[66]

When Buffon proposed fertile interbreeding in place of resemblance as a specific criterion, working naturalists reacted with impatience. Few naturalists, wrote Illiger, determined species from observation; "rather, they decide about species in their workrooms on the basis of certain diagnostic characters." Buffon's argument, that the physical reality of the species consisted in the constant succession of like individuals who reproduced one another, was given a standard interpretation: species characters were constant, and species members bred true. Buffon's suggestion, that neighboring groups in the table of organization, too distant to allow crossing, might descend from a single branch, could be read as an interpretation of the term *affinity*. What could be established without argument, of course, was the presence of structural likenesses and differences, however interpreted.

As naturalists continued their patient studies of affinity, they must have seemed impervious to ideas of physical descent and temporal transformation. That appearance was deceiving. The cornerstone of their practice, the essentialist species concept, which established the reality, constancy, and separateness of true species, simultaneously established the varietal concept as its necessary complement. In theory, the varieties absorbed the nonessential changes produced in the unchanging species by accidental causes—that is, climatic and edaphic conditions. Their influence was supposed to last only as long as they remained unchanged.

Many variations produced by external causes were ephemeral, but some proved to be more or less permanent. Did this mean, as Buffon seemed to argue, that new species might appear? Naturalists were aware of other situations that contradicted their equation of variation and external causes. Under certain circumstances, true species could fertilize one another and produce new kinds. Ongoing research demonstrated that these new kinds were not invariably sterile. What were the limits to the process? Gardeners and breeders crossed strains to produce new and constant races, and these in turn could produce secondary races. Was there a limit to the possible forms in some species? All of this evidence pointed to inner causes as a source of variation quite as important as external causes.

What kinds of inner causes? Obviously, these were causes that could generate and maintain—or alter—structure. Naturalists continued to regard species and outer form as one and the same thing, but in fact, as soon as they admitted inner causes for structural modifications, they initiated the separation of species from structure. The species became an agent of formation, whose effects, in conjunction with external influences, could be observed in the structure of the organism. This was an idea that occurred to several thinkers at about the same time but in response to very different problems. All of them, however, had participated in or followed the rejection of the doctrine of preformation in the science of generation, and it was no coincidence that these men were now reconsidering the constancy of the species.

In St. Petersburg, Caspar Friedrich Wolff observed changes in plants brought from Siberia. Over several generations, the plants changed gradually, and in the end they were identical with plants normally found in St. Petersburg. The species had not changed, only the plant's outer structure. Wolff concluded that species and structure were different things. The species, he wrote, "whatever it is, is the cause, whereas structure and form are the effect; the species itself escapes us and our sensations, while only the forms—its effects—are evident and observable."[67] The species consisted in the actual production of structure, but no aspect of that structure was essential to or characteristic of the species.

In Königsberg, Immanuel Kant traced the reality of the species to their common origin in a single lineage. Each lineage had a determinate number of germs and tendencies to be developed or repressed in response to external conditions. The structural differences that emerged over time were not specific differences, and they did not impair the unity of the lineage, which was based upon fertile reproduction, not upon resemblance.

In Göttingen, Johann Friedrich Blumenbach conceived a formative force that emerged from the organization of the reproductive fluid to produce, maintain, and transmit structure—always according to kind. Like any other force of nature, the formative force was subject to other physical influences. A change in normal conditions was liable to provoke a change in the formative force, which was then realized in a structural variation.

Although these three conceptions arose in very different contexts in response to very different problems, they resemble one another in important ways. All of them separate the process of formation from its results.

The results are observable in the outer and inner structure of the organism, but these results are no longer directly equated with the species; instead, they are equated with an underlying agent of continuity and change, whose presence can be inferred only from its observable effects. And as a direct result, the explanation of organic change is much more complicated. A simple variation produced by an accidental cause, for example, must first provoke a reaction in the agent of formation, which responds by producing an alteration better able to cope with the new condition. And these alterations occur in a determinate manner. Any transformation necessarily takes place on the basis of already existing qualities and forces, and this does not lead to wholly new species with entirely new characters.

The differences between these conceptions also deserve attention. Kant's lineage concept distributed natural forms along the axis of succession, where, as Buffon had argued, the physical repetitions and variations of forms were located. Blumenbach's formative force, on the other hand, arranged forms along the axis of simultaneity, where he and his contemporary systematists continued to locate relations among coexisting forms. From our perspective it is easy to see that there is no necessary equivalence between these two ordering principles. When Blumenbach began to consider Kant's genealogical method closely, he thought he could limit his acceptance to the atemporal, formal elements that harmonized with his own concerns.

Blumenbach's eclecticism was typical of those naturalists of the late eighteenth century who sought physical explanations for variation and anomaly. Buffon's theory of generation, suitably modified, promised a rational account not only of family resemblances but also of vices of conformation produced in the formative process. The genealogical method of Buffon (and Kant) was obviously serviceable with respect to human races, domestic animals, and cultivated plants. Physical descent was a major factor in the explanation of limits to crosses between true species. But these limited, individual considerations were not enough to bring about a change in the morphological criteria with which naturalists established the species. Nor could the idea of descent be extended above the level of the reproductively isolated species. At the very most, one can speak of an intensification of interest toward the end of the eighteenth century in viewing natural forms in terms of their history and physical descent.

IV

The Distribution of the Natural Kinds

Geographical history, to adopt the name given by E. A. W. Zimmermann, was the study of facts relative to the distribution of animals and plants over the surface of the earth and the general principles that naturalists deduced from those facts.[1] Obviously, geographical history was a forerunner of modern biogeography. But the discussion of facts and principles took place in a context very different from that of modern biology, and of course the context influenced not only the perception of facts but also the framing of problems and concepts. Prominent naturalists of the late eighteenth century—men like Johann Friedrich Blumenbach, Peter Simon Pallas, and Eberhard Zimmermann—repeatedly traced their basic assumptions concerning both geographical history and the related question of organic variability to two opposing traditions: the essentialism of Linnaeus and the Linnaeans and the more complex eclecticism of Buffon and Haller.[2]

In pre-Linnaean natural history, a highly developed conception dealt with the ultimate equilibrium among created beings. According to this conception, the order of nature was not only a vertical line, like a ladder, on which every created thing had a fixed place; nature was also active and dynamic. Her thronging creatures depended upon one another in the network of reciprocal relations that sustained her vital equilibrium. Whatever its diversity in the work of various naturalists and physicotheologians, the conception had at least one common thread: equilibrium was a principle that regulated natural objects, not an effect of their interaction.[3]

Linnaeus recast this conception in terms of analogies drawn from contemporary social theory, the economy, and the polity of nature. "The economy of nature," he wrote, "shows us the theology, or *fines rerum*, to what purpose each has been created and the combination that is among created things in their generation, preservation, and destruction."[4] Nature, like society, was a single hierarchical structure in which there existed a latent opposition between individual intentions and nature's impersonal requirements. For the individual, this meant struggle and death, but the general result was the equilibrium of nature. "The bloodbath and war of all against all, which at first seems so terrible, is not therefore something that ought to trouble us particularly. . . . Animals are created for a higher purpose, and since for this no great number of individuals is needed, it is reasonable that the superfluous should be eliminated."[5]

In the Linnaean conception of the natural economy, a single subordinate principle mitigated the rigor of this arrangement. Providence, Linnaeus argued, had divided plants and animals and confined them within well-established physical limits, the better to perform their tasks in the economy of nature. Moreover, the Creator had furnished each kind with a habit proper to the climate where it lived, so that it seemed destined solely for the place where it was found.

Naturalists experienced a great deal of difficulty in coming to terms with nature's spatial organization. Given the fixity of the species on the one hand and the recorded facts of distribution on the other, Linnaeus and the Linnaeans assumed that a design had preceded the deployment of forms, that distribution was subject to a plan. But did the plan unfold from a single center or scatter from many centers? By framing their questions in this way, by tying the function of distribution to its origin, naturalists were trying to think through a principle whose operation had brought about a puzzling state of affairs. "They sought," says Camille Limoges, "the regulated unfolding of natural order on the basis of a privileged moment, an ideal genesis, where all the rules were combined in a great knot."[6]

A well-known presentation of their predicament is found in Linnaeus's oration "De telluris habitabilis incremento," where Linnaeus revised Genesis to fit the known facts of organic distribution.[7] The idea that the earth had originally been of the same extent as it is today contradicted Scripture, assigned limits to creation, and found no support in the observable decrement of the sea. Linnaeus argued, therefore, that the land had steadily increased, and at the creation consisted of an equatorial mountain where every plant and animal found a proper station. As elevated parts of the globe appeared, the stations required by each kind recurred. Like physical conditions favored the growth of like kinds, and only migration and dissemination were needed to explain the present locations of species.

Scientific revisions of Scripture were commonplace at Uppsala in the 1740s, and Linnaeus's fellow academics—except, of course, the faculty of theology—read his argument not as an expression of heterodoxy but as the justification of an autonomous principle in the economy of nature.[8] Many elements in Linnaeus's presentation became standard topics in the discussions of geographical history. As late as 1812, competent naturalists were arguing, with only a few reservations, that primitive mountains were the main source of living forms, that the land was gaining upon the sea, that assemblages of living forms were determined by elevation and climate, and that like physical conditions everywhere proffered like kinds.[9]

When we come to Buffon's work after a prolonged study of Linnaeus, our first feeling is one of liberation. Buffon's advocacy of a descriptive, historical approach to nature, with his alternative theories of procedure, evidence, and truth, marked an important change in the conception of nature and the formation of concepts. The importance of the change is nowhere more obvious than in the intermittent discussion of endemism and degeneration in the *Histoire naturelle*. Buffon stressed the differences between quadrupeds in the Americas and the Old World and concluded that like physical conditions had not produced or favored like forms everywhere; he suggested that external circumstances had caused the differences in the species that lived in them. These ideas, like the contrasting ideas of Linnaeus, proved of some interest over the next fifty years.[10]

We should not, however, confuse these and related discussion with the whole of Buffon's theory of distribution. That theory evolved slowly over forty years, and in its final version, *Des époques de la nature*, was a bold and unorthodox statement of successive creation. The distribution of mammals, according to Buffon, originated in the polar regions. As the earth cooled, the poles first reached a temperature at which organized bodies could develop. There, the first animals—elephants, rhinoceroses, and others—were born. Nature had her original potency, so that these animals were larger and stronger than the related species that now inhabited the torrid zone. As the earth cooled, these animals moved toward the warm center: hence the carcasses found in Muscovy, Poland, and Canada. The heat of the earth decreased to such a degree that only its midsection retained enough warmth to accommodate these creatures. There they lived and would continue to live as long as the planet did not grow too cold. The oldest animals were followed by new animals that did not require an equal degree of heat. Again, these developed in the north and moved south as the cooling process continued. Finally, those animals appeared that still inhabit the polar regions—reindeer, polar bears, and others.[11]

Buffon published these conjectures in 1778, when the European tendency to interpret natural order as a single developmental process was already far advanced. His ideas were received with interest in circles outside the reach of orthodox science; contemporary naturalists, as always, took an independent line. What they found suggestive in Buffon's discussion had little to do with his so-called physical facts and hypothetical probabilities. Many of the elements in Buffon's speculation—his use of evidence drawn from fossils, stratification, and continental separation—

had already entered the European discussion of distribution. What interested naturalists, though, was Buffon's attempt to treat the problem of organic distribution as an independent historical process, self-sustaining and internally coherent.[12]

Twentieth-century specialists, too, have found Buffon's conjectures of interest. In their search for the origins of modern theories of variability, historicity, and distribution, historians of science have tended to concentrate on the relevant and immediately accessible statements found in Buffon and, to a lesser degree, on the dissertations of Linnaeus.[13] This is not to argue that Linnaeus and Buffon were not important or that their ideas were without influence. But all attempts to appropriate their general ideas directly have overlooked two factors of great importance in understanding the development of geographical history. Concentration upon theoretical statements has precluded any attention to the body of geographical fact collected and organized in the second half of the eighteenth century—a feat as impressive as, and not without relation to, the achievement of a natural method in systematics.

A second objection to the direct appropriation of eighteenth-century speculation is that it ignores the ambivalent attitude toward Linnaeus and Buffon in the early treatises on what was beginning to be called geographical history. The attitude is clearly seen in the first major work devoted to the subject, Eberhard Zimmermann's three-volume study, *Geographische Geschichte*.[14] Zimmermann repeatedly mentioned Linnaeus and Buffon with respect and acknowledged their mastery of the various fields bordering his discipline. Their opinions on the subject of distribution, however, he scrutinized so closely that he left none of their general ideas intact.

Zimmermann showed, for example, that the Linnaean concept of migration did not work. He conceded that, as the number of individuals in the original mountain garden increased, the pressure toward expansion grew. But how had plants and animals living in the elevated zones survived in the inhospitable lowlands and found new homes for themselves? The same climate everywhere did not bring forth the same forms, either. Linnaeus's conception of distribution was unworkable because it did not explain the main problem of geographical history, namely "the present cohabitation and subsequent partition within a country in and of itself on the basis of the present nature of animals."[15]

Buffon's theory of organic distribution, on the other hand, was illogical. The original animals ought to have spread from both poles to the equator, whereas not only was there not then living "any natural species from South America in Africa or South Asia, but this is also true of animals equidistant from the equator" (ibid., p. 198). Buffon did not improve his arguments by adding that the sea had altered the original situation, or that local temperatures were unfavorable, or that South America was suitable only for small animals. Buffon's theory was no more adequate than that of Linnaeus for comprehending the present state of distribution.

Zimmermann argued that geographical history would have to content itself in the immediate future with solutions to problems more modest than those proposed by Linnaeus and Buffon. The need was for more accurate, better-coordinated information on physical conditions and actual distribution on a worldwide scale. With the inchoate information naturalists possessed, they could only speculate aimlessly about possible migrations, adaptation to new habitations, and earlier geographical conditions. The *Geographische Geschichte* was an influential model for this more modest science. Many naturalists accepted Zimmermann's strategy; they avoided fruitless speculation and contented themselves with attainable goals. Much time was spent on the collection and analysis of material, and only toward the end of the century did they again attempt to articulate the material theoretically.

In the interval, the general outlines of a new empirical discipline had gradually emerged from a myriad of facts. When Gottfried Treviranus published the second and third volumes of his *Biologie* in 1803 and 1805, his discussion of organic distribution rested upon technical studies based upon a large and increasingly reliable body of information derived from analyses of the influence of physical conditions on organized bodies; upon studies of the physical nature of the locality in which the species and other groups customarily lived; and upon examinations of the geographical regions occupied by organized bodies on the surface of the earth. Naturalists were aware that the implications of these studies would result in new ideas, and they cautioned their readers against hasty generalizations. It was clear that their research rested upon reliable systematic characterization, but it was equally clear that the new discipline was not subordinate to systematics.

This is a tale that begins not with the work of an individual but with a scientific genre destined to multiply greatly during the latter half of the eighteenth century. Many countries, and almost all of the countries of Europe, often by province, canton, and even town, acquired for the first time a reliable flora, or a listing of their native plants.[16]

The exemplary flora contained a lengthy introduction describing the physical nature of the country and the general history of its vegetation. The history occasionally included a discussion of the relations of the vegetation with that of neighboring countries or even with the whole of the plant kingdom. Then followed a list of plants that grew in the country, arranged according to a method. The description of each plant included its name and specific character, a synonymy, a description sufficient to distinguish the plant, a detailed indication of the varieties found in the country, an enumeration of stations and habitations where the plant was found, and finally, an exposition of its local, practical uses.[17]

The table of contents of each flora shows how systematic and biogeographical information were, if not integrated, at least juxtaposed. The provincial flora was, in fact, an ideal location for kinds of information considered inappropriate to the great systematic works of the period, the *Species* and the *Genera plantarum*. Those works concerned types of organization and purely formal affinities. Naturalists regarded information about spatial and temporal distribution as circumstantial and secondary, although indispensable for practical knowledge of plant life. One important source for what was eventually to be called plant geography was this system of double-entry bookkeeping in the provincial floras.

I limit my consideration here to three floras, Linnaeus's *Flora lapponica*, Johann Georg Gmelin's *Flora sibirica*, and Albrecht von Haller's Swiss flora.[18] These works, compiled during the middle span of the century, served as models for European floras well into the next century and contained observations and generalizations that eventually became established topics in plant geography.[19] To trace the developing argument, I follow the order of their publication.

During his journey to Lapland in 1732, Linnaeus noted the elevations of different plant species. On this basis he distinguished in the *Flora lapponica* between the mountains and deserts of Lapland and laid out three mountain zones—low, mid, and high—and their respective vegetations. The high alpine region, he wrote, changed dress so "that I, who had thoroughly acquainted myself with Swedish plants, saw scarcely a Swedish plant in this Swedish land, but only strangers, plants never seen by me before."[20]

When he later consulted the work of other naturalists, Linnaeus found that many of these plants were common to other European mountains. "The greatest part of the plants," he wrote in his introduction to the *Flora lapponica*, "proffered by the remotest Alps thrive only there and are not found in the interjacent areas."[21] He went on to list plants common to the Lapland Alps and other European mountains. These were, he emphasized, the same plants, not to be found outside high mountain regions.

Although Linnaeus stressed elements common to the floras of widely separated regions, he could not ignore many differences. In the more southerly mountains of Dalarna, for example, although he did not find a single plant not previously seen in Lapland, the vegetation made a different general impression, and the dominant plants were quite different. He also noted that other plants appeared nowhere else. In mountain chains, he wrote, there are long stretches "that have some peculiar plants not sown in other very distant mountains and not found on interjacent peaks lying here and there."[22]

To explain these limited and heterogeneous observations, Linnaeus linked the dispersion of plants down mountains with their dissemination by water, wind, and other means and related these links to his basic systematic position, the origin of each kind in a single progenitorial unit. The result, given a final statement in his "De telluris habitabilis incremento," was a simple analogy drawn between external form and physical situation: like plants would be found in like stations and habitations.[23] Many, Linnaeus conceded, would find it difficult to conceive how all the plants on earth had been disseminated by single plants. He tried to show how from a single spot a tobacco plant might spread to every part of the world, and he concluded that any single plant could easily cover the face of the earth.

Naturalists did find it difficult to conceive how plants could have migrated from a single mountain in Asia Minor. In the *Hamburgische Magasin*, one of Haller's students at Göttingen, J. G. Zinn, argued that evidence drawn from alpine plants, far from indicating an original central location, undermined the idea altogether: "Everywhere those plants stand in our way which appear on the highest alpine peaks, where they could not have appeared, according to this opinion, unless they had previously grown in many lower regions, which, however, wholly contradicts the nature of these truly alpine plants."[24]

Plainly, Linnaeus's conception of migration was unsatisfactory. Alpine peaks and the Peruvian cordillera were separated from Asia Minor by hundreds of miles and broad oceans. "This intervening space is so great, the streaming ocean between so broad, that all conjecture falls to the ground: as if the seeds of these plants with the help of the wind or of the birds could have been brought from one precipice to another" (ibid.). The only possible explanation of this particular pattern of plant distribution was simultaneous creation and dissemination in different areas.

> For these reasons, it also seems highly probable that the Creator had brought forth at the creation of our earth many individuals of each kind, some in greater numbers, some in lesser, according to the different intentions whereby He had determined to distribute these plants over the whole earth and, according to the variousness of their structures and qualities, in appropriate areas and regions. (ibid., pp. 348–49)

Zinn concluded: "The more I regard all observations in respect to these questions that fall into our hands, the more I am convinced that quite different transformations must have taken place on the earth's surface than informed naturalists have heretofore considered" (ibid., p. 355). Zinn's remarks seem to have attracted no attention, and the article is perhaps best read as an indication of opinion at Göttingen.

A far more substantial criticism of Linnaeus's ideas was offered by Gmelin, a colleague in St. Petersburg, who hinted that religious preconceptions and limited research had dictated Linnaeus's results. In the preface to the *Flora sibirica*, Gmelin spent fifty-three pages on a physical description of the area of his investigation and then turned to the problem of distribution. He conceded that knowledge of the general laws according to which the Creator had distributed natural objects was worth any amount of labor, but those laws could be worked out only from the study of the present state of distribution.

Gmelin limited his consideration to European and Siberian vegetation, and he listed plants of various ranges under six general categories. In his second category, for example, he listed plants found in Europe but lacking in Siberia and plants growing in Siberia but rare in Europe. In his sixth category, he listed species found in both Siberian fields and European mountains. The patterns of distribution exhibited in these lists could not be explained on the basis of dissemination from a single, central location. They seemed rather to indicate that the Creator had dispersed some plants everywhere but had confined others within narrow limits, and these had spread only to neighboring regions. Assuming the permanence of the species, the Creator must have sown plants of the same species in different regions.[25] Although tentative and based on limited evidence, Gmelin's statement provided a nucleus for the concept of centers of creation.

A long and enthusiastic review in the *Göttingsche Anzeigen* was silent about Gmelin's conjecture about simultaneous creation. The reviewer was polite about Gmelin's careful treatment of relations between Siberian and European vegetation, but only Gmelin's sixth category was considered to have mustered enough evidence to establish itself scientifically. "The assertion that alpine plants of warmer countries grow in the fields of colder countries has been entirely confirmed by experience in Siberia and Kamtschatka, where our Swiss alpine plants grow in the meadows."[26]

Linnaeus, too, noted the solidity of Gmelin's sixth category, and he returned to the problem of alpine vegetation in his dissertation "Flora alpina." Before the publication of Gmelin's work, Linnaeus's knowledge of Siberian plants had been meager; he had assumed that the presence of alpine plants indicated mountain regions. He did not abandon the idea altogether, but he did modify it to accommodate Gmelin's evidence. Alps, Linnaeus wrote, were found everywhere in the world, even in the tropics; but

> the warmer the tracts in which mountains are situated, the higher these must be in order to assume an alpine nature, and the reverse—the nearer the poles, the lower they may be in order to have such a nature. The greater part of Siberia is of alpine nature, as clearly appears from Gmelin's Siberian flora. [He added in 1759: In just such a way, Greenland and Spitzbergen bring forth purely alpine plants.] Therefore all Alps ought to be regarded, however distant from one another, as one and the same nature spread here and there around the globe.[27]

When Albrecht von Haller returned to the problem of alpine flora in the 1760s, he noted new evidence from Greenland and Spitzbergen that supported Gmelin's sixth category, and he called attention to the fact that alpine plants occurred there in coastal regions and went on to correct Linnaeus's formulation. Returning to the mountain regions laid out in his own *Enumeratio*, Haller sketched the physical characteristics of seven Swiss mountain regions and their dominant plants.[28] In the third region, alpine pastures, he noted the occurrence of numerous plants "quae vulgo alpinae dicuntur." The same plants were found in Lapland, Siberia, and Kamtschatka; in Switzerland only the peaks of mountains exhibited these plants. Haller was content to state these facts—he did not attempt to explain them.[29]

That this was a deliberate strategy, understood among specialists, is shown by the best research on distribution in the period immediately following the Swiss flora. Many contemporary naturalists followed Haller's example and limited themselves to careful enumeration of the plant stations and habitations in various areas of the world. Not all naturalists formed their collections in the same way or even based them on a single system of classification, and their terminology tended to change with the needs of their special projects. One problem confronted by each of these men was the need for a reference system (a method for arranging entries) and for descriptive models upon which the entries could be built.

Perhaps the best way to come to terms with the resulting complexity in distributional studies during the later eighteenth century is to examine three typical contributions. I have chosen Peter Simon Pallas's analysis of the regions of Siberian vegetation in 1772, work that grew out of the collections of Johann Reinhold Forster and his son Georg in the South Pacific between 1772 and 1775, and Ramond de Carbonnière's studies of vegetation in the Pyrenees in 1787. The choices are somewhat arbitrary but are dictated in part by the importance of the studies for later work on distribution. They constitute a very small sample of the work actually published in the period, even for the limited areas I have chosen. In Russia, for example, in addition to the travels of Pallas, important accounts of natural objects and ethnography were written by S. G. Gmelin, J. G. Georgi, J. A. Güldenstädt, Iwan Lepechin, and Freyherr Marschall von Bieberstein. In the south Pacific, the Forsters were preceded by Philibert Commerson, Joseph Banks, and Daniel Solander. For the French mountain regions there are important works by H. B. de Saussure and J.-L. Giraud-Soulavie. And of course other areas of the world—Asia, Africa, and the Americas—were explored with equal intensity. A complete treatment of contributions to botanical geography at this time, even if limited to those made by northern Europeans, would require another book.

During the 1760s, the Academy of Science in St. Petersburg undertook a concerted exploration of a region extending from the polar sea to Persia and from the Caucasus to the upper Amur. The central region in the academy's planned investigation included the upper Urals and western Siberia, the Altai and Sajan mountains, the areas around Lake Baikal, and the upper Amur. The region was allotted to Peter Simon Pallas, who directed the entire exploration.

Pallas drew on the work of many predecessors, including J. G. Gmelin. In his description of the physical nature of Siberia, Gmelin had noted the appearance of an Asian flora east of the Jenissei. By August 1772, Pallas had explored the area and had divided the region into six floral provinces on the basis of topography. The first change in vegetation, he wrote, began east of the Urals, where a Pannonian flora was dominant. From Irtisch toward the foot of the Altaic Mountains the change was even more marked, while the mountains offered a flora comparable to that of the Jenissei. On the heights beyond the Ob grew plants found, to the west, only in the Altaic Mountains. Above the Jenissei were many mountain plants, some native but some common to a district south of the Baikal, which seemed to be their point of origin. The lower mountain area between the Jenissei and Baikal, however, was dominated by a cold forest and meadow flora. The mountain region around Baikal offered the most rare plants, some growing in warm, open fields and others growing on snowy peaks and in cold valleys. Finally, high alpine plants were found in eastern Siberia and Kamtschatka, on low mountains, plains, and in swamps.[30]

I have simplified Pallas's results, but I have not otherwise been unfaithful to his impressionistic presentation. His party did not include a surveyor, and Pallas was not able to indicate elevations, latitude, or longitude.[31] The plant lists that Pallas appended to these topographical regions were a more important aspect of his work. The lists were not complete; they contained only what Pallas had collected in a year's time. Still, Lamarck's complaint—that Pallas had discovered few plants and no new genera—was wide of the mark. Pallas was concerned not with completing the systematic investigation of formal types but with the physical principles underlying the differing distributions of plant groups. His lists were a preliminary sketch for a general disposition of plants on the soil of Russia projected for the conclusion of the *Flora rossica*, not unlike the botanical map of France eventually found in *Flore française*.[32]

French investigators of mountain zones tended to focus on quantifiable physical factors. Basing their work on the study of plants in widely separated regions, they were aware of the associations of plants in communities, but their interest in these phenomena was perfunctory. In this respect, Ramond's account of vegetation in the Pyrenees is representative of the direction of French research.

In the course of a journey to the Pyrenees in 1787, Ramond analyzed the consequences of differences in elevation and climate for plant life. Mountains, taken from base to summit, presented a disposition of plants parallel to that exhibited by the surface of the earth extending from the site of the mountain to the pole. The scale of vegetation extended from the snow line downward, modified slightly by subordinate conditions such as light and air. Thus, Ramond concluded, "the disposition of plants on the slopes of mountains answers mainly to the temperature of the different zones."[33] Temperature was relative, of course, not only to elevation above sea level but to latitude.

At the lower limit of the zone of ice Ramond found an arid zone where a few mosses took advantage of every surface extracted by local warmth to push up to their legitimate term (ibid., p. 332). Immediately below the mosses Ramond located an alpine flora proper, "a certain number of perfect plants of a particular character, wild . . . and voluntary, suited only to these heights of which no treatment may deprive them and no care naturalize elsewhere" (ibid., p. 333). Ramond placed the first shrub, a rhododendron, at about 300 to 350 fathoms below the snow line—that is, in the Alps at about 800 fathoms elevation and in the Pyrenees at 900 fathoms. Finally, slightly below the zone of shrubs, the first trees appeared: the *if*, the *pin cembro*, and the *pin sauvage*. "Sage nature," Ramond exclaimed, "in her plan, in that plan as admirable for its simplicity as for its extent, all is analogy, everything repeats itself under all possible forms" (ibid., pp. 336–37).

As the remark indicates, Ramond, like Linnaeus, assumed that like plants grew in like situations, an assumption that limited his comparisons to common elements in the vegetation of the Alps and the Pyrenees. Within these bounds, the assumption was not inappropriate. The mountain regions of France offer a number of plants peculiar to themselves, most of which are found in all of these mountains. "For whatever differences," Candolle wrote, "are presented by the chains of the Vosges, the

Jura, the Alps, the mountains of the Auvergne, of Cevennes, and of the Pyrenees, we must admit that the aspect of their vegetation offers many traits of resemblance and that the majority of mountain plants are to be found in these different chains."[34]

These reassuring facts, and the assumptions to which they gave rise, were not available to the many naturalists at work outside of Europe. These men continued to think of their work as a contribution to the inventory of nature, and upon their return to Europe they willingly limited themselves to the description of new genera and species. But because the dominant Linnaean system, which was matchless as a sorting device, included a note on station and habitation in the generic and specific characters, the lists of new plants did contain some geographical information.

A volume published by Johann and Georg Forster shortly after their voyage with Captain James Cook, *Characteres generum plantarum*, is an example of this conservative systematic research. The Forsters worked within the boundaries of Linnaean natural history and identified their science with the collection, analysis, and classification of natural forms. Their genera were laid out according to Linnaean descriptive models and ordered according to the Linnaean sexual system. Georg gave a brief description of their results, published in that organ of Linnaean science, the *Kungliga svenska vetenskapsakademiens handlingar*.

> When my father and I returned from our voyage around the world, we carried with us about two hundred and fifty new species of plants previously unknown to the learned world. We established about seventy new genera, which contained approximately a hundred new species, but the rest of the plants we brought home were classified according to genera already established by the Archiater von Linné.[35]

The learned world viewed the Forsters' work as a contribution to the inventory of nature in the Linnaean manner. Great expectations had formed around the accessions of the two naturalists, but these were dashed when parts of the harvest were published. The sages murmured at how little had been found. The older Forster conceded in his preface to the *Characteres* that seventy-five new genera seemed a poor return; perhaps he had erred on the side of caution. His reader had to remember that half of the voyage had taken place during the autumn and winter, that long stretches were spent aboard ship, and that in areas where much was expected, little was found.[36]

The hostility of the learned world was absurdly exaggerated. The sobriety and restraint of the Forsters' systematic work proved to be its strength, a strength that had little to do with the number of their accessions. As in the case of Pallas and Gmelin, it was the scrupulous plant lists

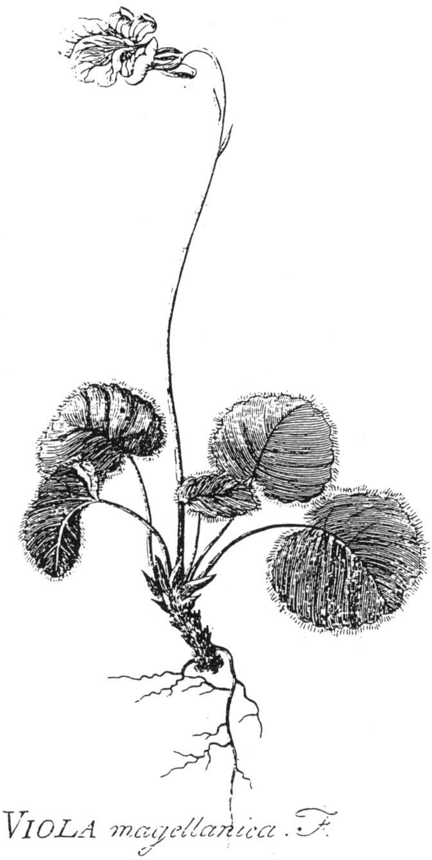

Georg Forster's drawing of the Viola magellanica F., *an accession from the voyage with Captain Cook*

and the careful record of stations and habitations that provided solid evidence for unsuspected patterns of distribution.

In his *Observations Made During a Voyage Round the World,* the older Forster fit his observations into a Linnaean framework. Distribution presented a unified pattern based essentially upon temperature. Vegetation, from the luxuriance of the tropics to the scanty life of the polar zones, was "in proportion as places lie nearer the sun."[37] And as one sailed through different regions, vegetation in each new station partook of both the preceding and the succeeding region in direct proportion to proximity. There were exceptions to these two rules: a few plants were common to all climates of the south Pacific; other species obviated differences in climate by a higher or lower elevation; and finally, a similarity of situation and climate sometimes produced a similarity of vegetation: "this is the reason why the cold mountains of the Tierra del Fuego produce several plants that in Europe are the inhabitants of Lapland, the Pyrenees, and the Alps" (ibid., p. 176).

When Georg Forster returned to this piece of evidence ten years later, he rejected the Linnaean commonplace, noting the presence of *Pinguicula alpina*, *Galium Aparine*, *Statice Armeria*, and *Ranunculus Lapponicus* in both Lapland and the Tierra del Fuego. He thought that neither location could be considered the unique point of origin for these plants of the same species. It was more probable that the surface of the earth had produced many individuals of each species simultaneously and that plants appeared in similar or dissimilar form, all according to climate.[38]

This sampling of the early contributions to botanical geography demonstrates how fragmentary and even chaotic the subject could be. What is less obvious is that each of these naturalists, in his own way, distinguished between the physical locality, the climatic and edaphic conditions in which species grew, and the geographical regions where they were found. The broad conclusions drawn from the resulting patterns of distribution began to be published in the German-speaking world after 1790.

One of the first syntheses was written by K. L. Willdenow of Berlin. Willdenow's effort amounted to no more than a subordinate chapter in an elementary text; his ideas were influential, however, because they were integrated into a single argument. He showed how the study of physical conditions, plant stations, and plant habitations built upon one another in logical progression to provide a framework for geographical history. He also coordinated physical geography and the general character of vegetation to produce five main European floras. This was an extension of Pallas's work on the Siberian regions but required disciplined knowledge of a greater range of materials. The novel element in Willdenow's argument was his insistence that many features of present distribution were products of historical development. Here it is natural to want to see the influence of Zimmermann's earlier work on quadrupeds, but Nils von Hofsten has pointed out that both botany and zoology possessed vast collections of forms and observations, and investigators with parallel methods were inevitably forced along parallel paths.[39]

In the first edition of the *Grundriss der Kräuterkunde* (1792), Willdenow treated plant geography in Linnaean terms. The distribution of plants over the surface of the earth could be explained on the basis of climate and migration. In later editions, Willdenow continued to argue for Linnaeus's hypothesis that the original mountains of the planet were the chief source of regional floras. Willdenow's use of the plural *mountains* was a discreet rejection of Linnaeus's single, original mountain. Willdenow continued to find mountains particularly efficacious points of departure for plants, and he proposed multiple regions of creation centered on primitive mountain ranges.

As we have seen, naturalists had difficulty explaining patterns of distribution analogous to those Gmelin had found in Siberia. Given, on the one hand, the almost universal belief in the act of creation and the fixity of the species and, on the other hand, the actual patterns of distribution, naturalists could not explain why some species were confined to a single narrow region rather than being evenly distributed about the earth or why other species were found in widely separated regions. Willdenow, like other German naturalists before him, proposed a world in which distribution had been ordained by the Creator. In the beginning, God had either disseminated or—in Wilddenow's opinion—created plant kinds in the regions they now occupied; and once the individual kinds had been placed in those regions, each had spread as far as its own nature with respect to climate allowed.

Further, each region had undergone gross physical changes, and these changes too had affected distribution. Thus the composition of the present botanical regions was, to a degree, a product of migration and physical history. Accordingly, Willdenow redefined his subject to include "the changes which plants have most probably undergone during the various revolutions this earth has suffered."[40] He argued that the work of Cuvier, Batsch, Blumenbach, and Lichtenberg showed that at least one creation had come to naught and that the present organic world was a new appearance. Several thousands of years, if not more, had elapsed before all things had come to the state in which they currently existed. Areas separated by oceans presented similar floras; in former times, these areas were most probably joined. "At least we may suspect this from the different natural productions that both have in common. Thus New Holland may have been joined to the Cape of Good Hope and Norfolk Island with New Zealand" (ibid., p. 485). The occurrence of seashore plants near salt springs seemed to indicate a period when the seas were more extensive than they presently were. The alpine floras common to Siberia and Lapland and to the Pyrenees, Apennines, and Carpathians proved that these mountains had cohered, although winds, birds, and other circumstances may have contributed (ibid., pp. 489–90).

Comprehensive as these explanations were, they could not be made to cover all of the discontinuities known to Willdenow. He was particularly distressed by a piece of evidence from the Forsters' description of the Tierra del Fuego. How to explain the appearance there of northern species? Because physical history provided no plausible explanation, Willdenow questioned the Forsters' systematic abilities. "Perhaps the great likeness between the European and southern plants misled these great philosophers, though there might be distinguishing marks, which, however, the two gentlemen, firmly believing them to be our European species, did not attend to" (ibid., p. 608). He retreated to the standard Linnaean explanation, that like situations produced like plants. "And why should nature not produce, under different latitudes and longitudes, species which are very like each other?" (ibid., p. 609).

This willful distortion is a minor blemish in Willdenow's presentation. The merit of his work, like Zimmermann's, is that it applies the historical rationale consistently, if not always convincingly, to intractable problems in the study of distribution. By doing so, Willdenow established historical considerations as a factor in plant distribution almost as important as climatic and edaphic conditions. Although he was not the first to conceive plant distribution as a product of historical development, Willdenow did establish the development as an integral part of plant geography.

In the ferment of German scientific life at the time, it was not equally obvious to all that history had become a cultural leitmotiv. Treviranus, for example, used the word "process" on many pages of his plant geography and professed to believe that similarities in floras of areas then widely separated indicated an earlier connection. But the impression left by his botanical geography is ahistorical. A reader does not have the sense, as he so often does in reading Willdenow, that the plant world was a product of great revolutions in former stages of the earth. This is in part because of Treviranus's ambition to give a purely formal structure to plant geography that was not merely national or European but included the entire surface of the earth. His argument built upon the familiar triad of physical agents, plant stations, and plant habitations but extended and generalized the analysis to produce eight main world floras. As in the work of his predecessors on this subject, Treviranus's approach was purely analytical and synchronic.[41]

Even where Treviranus might have used historical explanations he was concerned with another kind of argument. The basis of his biological synthesis was a dynamic conception of nature based upon the interaction of formative forces and formless matter.[42] All organisms, Treviranus argued, issued from formless matter, and formative forces have brought forth autochthonous individuals everywhere. In places where the climate and other physical circumstances were similar, these autochthonous individuals would be alike, and the species that developed from them would everywhere remain alike as long as the influences they underwent did not change. Similarities in floras and faunas of widely separated areas were not difficult to explain—the same physical conditions produced the same plants. On the contrary, it was the differences among floras of areas with similar climates that constituted the real problem. Treviranus suggested the influence of cosmic galvanism.

And it is perhaps for this reason that the naturalist often described as the founding father of plant geography, Alexander von Humboldt, was so circumspect in his discussion of this speculation in his *Essai sur la géographie des plantes*. Like the greatest of his predecessors, Humboldt preferred attainable goals, and he seems almost to have convinced himself that plant geography would have fulfilled its task when it had laid bare the laws governing the present distribution of plants. The operative word here is *almost*: no serious naturalist, certainly not Humboldt, could have ignored the inroads being made upon natural history by historicism and *Naturphilosophie*.

Humboldt took up many of his friend Willdenow's historical aperçus and developed them with the curious blend of facts and conjecture that characterizes so much of his work. A trained geologist, he appreciated Willdenow's suggestion that the study of paleontology bore upon plant geography. Petrified fruits, palms, and ferns buried in the glacial earth of the north indicated a climate once capable of supporting these creatures. Not inconceivably, the earth in passing from a liquid to a solid state had liberated a tremendous amount of caloric, increasing regional temperatures independently of solar temperatures.

As for subsequent transformations of the earth, the habitations of plants showed that the catastrophe that had opened the Straits of Gibraltar and hindered the spread of African plants had no parallel in the Americas, where oak species on the heights above Tenochitlan were identical with those at a latitude of forty-five degrees. As for the liaison of continents, plant geography would demonstrate that islands now separate, and even continents like Africa and South America, had been joined before organized beings had developed. Humboldt also mentioned the evidence provided by garden and crop plants, which, serving as hinges between human societies and their biological environment, revealed the sequences by which the human species had spread over the surface of the earth. Not one of these insights was developed for more than a page, and Humboldt apparently regarded them as no more than suggestions for new research.

Another influential part of the *Essai* concerned the graphic representation of plant groups. As we have seen, studies of physical conditions and their effects on vegetation and its distribution were well established by the middle of the eighteenth century and continued to develop over the next fifty years. Ramond, Candolle, and others had noted the altitudes for zones of vegetation in the Pyrenees, the Alps, and other mountain chains, and Humboldt made ample use of their research. But by utilizing instruments to measure not only altitude but temperature and air pressure as well, he correlated plant zones and physical factors with greater precision. Humboldt was equally interested in an accurate representation of horizontal ranges. Again following Willdenow, he distinguished between species that occurred as isolated individuals and social species that dominated wide areas. This constituted a major contrast between the plant communities of the temperate zone and the diversified assemblages of the tropics. Humboldt, joining this idea with an old technique of J.-L. Giraud-Soulavie's, proposed a map of these associations. For horizontal ranges the data for the Americas at this time were too scanty, but by introducing profiles of mountain ranges with the limits of vegetation indicated, Humboldt created a graphic tool of great potential.

With this new precision, the old problems relating to plant habitations became increasingly importunate. Humboldt was aware of the discontinuous ranges of alpine and tropical plants and of analogies among widely separated floras. He was disturbed by both situations. Fieldwork had convinced him that similar climates do not produce or favor similar plants. The uniformity of tropical floras characterized only coastal regions, where plants had become pantropic through human activity. In inland regions the same plant species—with the exception of some species of cryptogams—did not occur naturally in both the Old World and the New. The wild strawberry, for example, both in Canada and the United States differed from the European species, and Humboldt wrote that, if he had found the plant in the Andes, it too would have differed specifically from *Fragaria vesca*. In fact, "during the five years we botanized in the two hemispheres, we did not collect a single European plant produced spontaneously by the soil of meridional America."[43]

How did Humboldt explain the presence of related but not identical kinds in the Old and the New Worlds? Two answers are possible, neither of them quite right. At one point, Humboldt wrote that the diversity of species in the Old World and the New was the effect of "a degeneration that, in time, has rendered constant varieties at first accidental."[44] I should like to read this as a personal reflection on the vague evolutionary ideas of the period, but I cannot legitimately do so. The statement was an echo of Buffon by way of Zimmermann, and Humboldt did not develop the idea further.[45]

Another possible explanation is the idea of centers of creation. In the *Essai*, Humboldt claimed that only the germs of cryptogams had developed spontaneously in all climates. Later he stated expressly that the same phanerogams had appeared in both America and Europe, in both the northern and southern hemispheres. He rejected the Linnaean theory of migration. The more we study the divisions of organized beings, he wrote, "the more we are inclined, if not to renounce ideas of migration, at least to regard them as not entirely satisfying hypotheses."[46] All of this points toward a conception of centers of creation not so very different from that of Treviranus. But in combating hypotheses too easily adopted, Humboldt did not mean to substitute others only slightly more satisfactory. "The causes which have restricted each species in more or less narrow limits are covered with that impenetrable veil that hides from our eyes everything that relates to the origin of things" (ibid.).

When we approach Humboldt's *Essai* through the work of his predecessors, it is obvious that his contribution to plant geography did not consist in the originality of his views but in the ways in which he coordinated observations (his own and others) with more general ideas. Humboldt understood this very well; he preferred, he said, "the connection of facts which have long been observed, to the knowledge of insulated facts, although they were new."[47] Once he made these connections, it was clear that many of the previous century's stock theories of plant distribution would have to be discarded. The Linnaean hypothesis of a single point of origin did not work: migration could not explain all of the present habitations of plants. The alternative hypothesis of a simultaneous creation in or distribution to many places was scarcely more satisfactory: mountains were not necessarily the sources of regional floras; more often, they served as barriers. Similar climates obviously did not produce or favor similar plant kinds.[48]

In fact, among the general phenomena presented by the habitations of plants, there were three that, given the fixity of organized beings, could not be reduced to a single theory. Species of some genera and families grew in a single region; thus, all of the species of the genera *Mutisia*, *Cinchona*, *Fuchsia*, *Cactus*, and *Tillandsia* originated in equatorial America. On the other hand, some genera contained species that grew spontaneously everywhere—mosses and lichens, chiefly. Finally, certain regions offered no like species yet gave birth to analogous species belonging to the same genera; thus, different species of the genera *Fraxinus*, *Populus*, *Pinus*, and *Tilia* were divided between the United States and Europe. This irregular disposition of species and genera was well established and generally accepted, and later evidence confirmed it. The causes that restricted some species or produced constant varieties could be worked out in part, but only in part, from physical or physiological conditions. Other causes, which seemed to issue from the origin of things, had to be considered incomprehensible according to assumptions that Humboldt shared with many of his contemporaries.[49]

During the final third of the eighteenth century, the discussion of plant distribution trailed the parallel study of animals in one important respect. Botanists continued to describe plant distribution as wholly dependent on climatic and edaphic conditions and stated repeatedly that countries with the same physical conditions would harbor the same plants. Eberhard Zimmermann was simply repeating learned opinion when he wrote in 1778 "that one and the same physical climate favors or produces the same plants."[50]

Zimmermann understood that he was dealing here with a general problem that he would solve in a very different way with respect to animals. On the basis of reports available to him, however, he could only state that the reason one did not find all the plants of Greenland in the Swiss Alps was because secondary causes, such as soil, light, and temperature, were not exactly the same in Greenland and the Alps. The evidence with respect to animals was not markedly better, but it did show that the general ideas about distribution then current did not and could not explain or predict the many anomalies found in the animal kingdom.

Zimmermann's study of the distributional patterns of mammals comprised three distinct elements. The first of these was an analysis of the general physical and historical determinants underlying observable patterns of distribution. A second element consisted in an inventory of the habitations of all known mammals. The third element was a theoretical discussion of the system or plan of distribution derived from the studies of physical determinants and the actual patterns of distribution. In other words, Zimmermann's argument built on the logical progression from physical conditions to stations and habitations that we have already identified in the parallel study of plants.

Climate was, of course, the simplest and most general of physical determinants. Because animals, unlike plants, were both mobile and highly differentiated, Zimmermann concentrated on the correlation between structure and climate. Only a few species—humans and dogs, for example—were so constructed that they defied all climates; most animals were less well constructed and were therefore confined within climatic limits of greater or lesser extent. The structure of their bodies was, in terms of teleology, adapted to a particular way of life, and they seemed destined solely for the places where they were found. The naturalist might try to establish some further correlation between the strength, endurance, or systematic affinities of a species and its distribution, but, warned Zimmermann, he would find no single, simple solution to his quandaries (ibid., 1:20–21).

The second major determinant of distribution discussed by Zimmermann was the physical history of the earth. Mammals confined to a single parallel found the oceans a barrier to their expansion. Naturalists could plausibly argue, therefore, that when two separate regions had different faunas, the separation had in all probability always existed, and conversely, that if two now-separate regions had similar faunas, these regions had previously been one (ibid., 1:29–30).

At this point, Zimmermann introduced the question of variations caused by external physical conditions. The existence of these variations had long been established, at least among domestic species, and some were known to be hereditary. This did not imply, however, that physical conditions could produce permanent variations or new species. Zimmermann further stipulated that two animals from different climates should be treated as separate kinds, although they might appear to be the same kind. Every kind had a place and had been "distributed over the earth from the very beginning, each situated in a climate appropriate to itself" (ibid., 3:192; see also 1:23–24, 25). Hofsten has remarked that this assertion never attained the dignity of a rational principle in Zimmermann's work; it was simply a basic assumption, to which Zimmermann added a few qualifications. Zimmermann did admit that some kinds had been lost in the wake of catastrophes and that wholly different kinds had been formed; he also thought it likely that in similar situations some species had approached others so closely that their crosses had brought forth new kinds (1:26, 27).

The general distribution of mammals, then, was determined by physical conditions and by the history of the earth. Yet this truth could not be used to justify Linnaeus's facile generalization that like or comparable climates everywhere harbored like species. "In that case," said Zimmermann, "animals from the Cape of Good Hope or some parts of South America would be found in Morocco, Spain, and Egypt" (ibid., 1:28). And this was not the case; South America harbored its own peculiar faunas, and so did the Cape. "Here a distribution of quadrupeds has preceded the beginning," Zimmermann claimed, "and it is difficult to think of the expansion of these animals from one and the same land" (ibid.). Admittedly, such a state of affairs could not be reconciled with the tale found in Genesis, but an enlightened age no longer expected this.

Nor did Zimmermann believe that Buffon was justified in accounting for the endemism of New World faunas by proposing the degeneration of Old World forms in the enfeebling conditions of a new continent. Obviously, physical conditions caused physical changes, and a change in situation or nourishment was apt to bring about a variation in the specific form, but the degeneration of wild animals through climate offered far more difficulties than Buffon appeared to think (ibid., 1:27). Naturalists were not justified in concluding, on the basis of changes brought about by climate, that "wholly new species had appeared." This, Zimmermann argued, was an entirely new question, not answerable in the lifetime of a single observer (1:25). Zimmermann defined his task as the representation and explanation of the distribution of quadrupeds over the surface of the earth, that is, their cohabitation and partition of its various regions on the basis of their present natures. His method was analogous to that of the systematists. "Just as the enormous multitude of entities in nature follows a plan," so he hoped "to be able to discover in the manifold and division of these things a similar order" (1:7).

More than half of Zimmermann's three-volume work was devoted to an exhaustive inventory of the quadrupeds then known. In relation to each kind, he discussed the correlation between the structure of the animal and the extent of its distribution, the effect of physical conditions on the nature of the animal, and the ways in which these effects pointed to a probable point of origin. Nothing Zimmermann wrote is more difficult to read today with patience than this inventory, but for his contemporaries it may have been the influential section of his treatise; it is a complete repository of fact and opinion concerning the distribution of mammals as understood in the 1780s. Zimmermann himself regarded this catalog as the indispensable preliminary to the general discussion in his final volume.

There Zimmermann proposed five major regions—Europe, Asia, Africa, America, and the larger islands of the South Pacific—and in these regions he listed the quadrupeds known to him. He was particularly careful to indicate kinds common to the Old and New World and to mention the fatherlands or limits of some families (ibid., 3:90). He presented the same information on a geographical map, with the names of animals printed over their proper locations and the limits of the major regions drawn emphatically in different colors.

He then turned to the question of whether quadrupeds had been distributed in this way since the beginning of time or whether they had moved to their present locations from some originally limited area. The question, he insisted, could be answered only on the basis of currently known facts and the currently established course of nature. His inventory established that only a few animals were so constructed that they could endure all climates; a larger number were enclosed within narrower limits; and the vast majority were restricted to very small areas. The chief cause, also established by the inventory, lay in the nature of the animals themselves, in the limits to their endurance and their malleability (ibid., 3 : 189). Research had further established that crosses between families did not result in fruitful offspring. Nature had drawn firm lines around the animal families, and neither human art nor the influence of climate could break those lines (3 : 190).

On this basis, that is, the present facts of distribution and the unchanging laws of nature, the naturalist could affirm only that "the animals had been distributed over the earth from the very beginning, each situated in a climate appropriate to itself" (ibid., 3 : 191). For if he argued otherwise, "then the previously cited, currently valid ordinations of nature are false" (3 : 189). One could not imagine the reindeer and the lion living together: neither could tolerate the other's climate. To imagine them coexisting was to imagine their natures different. "When then I work from the true present facts backwards, then an original distribution of animals, each on a part of the earth's surface suitable to itself, is the most reasonable assumption, and one that agrees both with our observations and with a wise arrangement" (3 : 203).

The argument against a single, original homeland for the quadrupeds also affected Zimmermann's conception of migration. The quadrupeds had appeared here and there in the beginning, the carnivores somewhat fewer in number than the herbivores. They had reproduced and spread as far as their natures, with respect to climate, allowed. In other words, natural migration was based on population pressures. Following Linnaeus, Zimmermann argued that the equilibrium among specific populations, also established from the beginning, had remained constant. Migration was triggered as well by a number of circumstantial factors, such as want of nourishment, climatic change, human interference, and revolutions of the earth's surface. For each factor, Zimmermann diligently adduced a list of examples, but his historical bias led him to amplify the discussion of the fourth factor, the correlation between animal migration and the history of the earth.

Inasmuch as the present form of the earth's surface had obviously changed from past forms, through both fire and water, animal habitations had necessarily changed as well. The evidence derived from these habitations, when read aright, would elucidate the history of the earth. It was clear that, "when closely situated countries, or countries with like temperatures under one and the same heaven, contained like or unlike animals, then one could conclude not improbably a previous connection or separation" (ibid., 3 : 216).

The present animal populations of many islands (Corsica, Great Britain, Madagascar, and Ceylon) indicated an earlier union with adjacent continents. Similarly, a naturalist could conclude that Europe and Africa, North

America and northern Asia, had once been joined. But it was also beyond question—as long as one did not accept Buffon's theory of degeneration—that America had not cohered with Africa since the appearance of animal populations. The evidence from Zimmermann's inventory was so overwhelming on this point that, even if a few species turned out to be common to the tropical regions of both worlds, this would only prove "that in both worlds, under like climates, a pair of each animal kind has been brought forth" (ibid., 3:239). But it would not explain why all of the other kinds in the two worlds were so totally different.

My summary schematizes Zimmermann's position on the problem of distribution. I have tried to retain as much as possible of Zimmermann's original language, his historical framework, and his views of other theories. In none of these areas was he profoundly original: his terminology was common coin in the later eighteenth century, when many naturalists applied historical explanations to individual distributional phenomena; and the theories of both Linnaeus and Buffon were more complex and enigmatic than Zimmermann represented them. Yet the three volumes of the *Geographische Geschichte* were an important contribution to natural history. Given the state of informed opinion on the problem of distribution and the well-nigh universal fixist conception of natural forms, Zimmermann gave as consistent and thorough an interpretation of the phenomenon of animal distribution as his period allowed.[51]

It is important, therefore, to keep in mind the chief ideas that served as points of reference for subsequent research. Like most of his contemporaries, Zimmermann continued to accept the idea of an original creation and the permanence of the species. But given the patterns of distribution of the animal kinds, he found it considerably more difficult to accept the mutually contradictory hypotheses proposed by Linnaeus and Buffon. Linnaeus's theory of migration from a single location could not explain adequately why some species were confined to a single narrow region rather than being more evenly distributed over the earth or why other species were now found in two or more widely separated regions. On the other hand, Buffon's theory that the quadrupeds had moved from the pole in the course of a cooling process was illogical and did not represent known facts, while his use of the concept of degeneration to explain the endemism of New World faunas conflicted with the view that the natural kinds were unchanging entities created by God.

To get around these difficulties, Zimmermann proposed (and he was certainly not the first to do so) an ordered world in which distribution had also been planned and executed by God. According to this conception, God had created the living kinds within the regions they now occupied; the world since the beginning had been divided into these regions. And once the kinds had been placed in these regions, each had spread as far as its own nature, with respect to climate, allowed. Moreover, each region had undergone gross physical changes in the course of time, and these changes had also affected the distribution of the natural species. Thus the composition of the present regions was a product of migrations and physical history, at least to a degree. A naturalist mapping these different regions and interpreting their historical evidence might eventually hope to work out the original plan of distribution, just as the systematist in his analogous situation, cataloguing formal affinities, hoped to reveal the formal plan of creation.

The conception of distributional regions, as we find it in the *Geographische Geschichte*, confronted Zimmermann and the other German naturalists who accepted these ideas with intractable problems for which they could not find convincing solutions. The first problem had to do with the delineation of regions of distribution. In Zimmermann's view, there were five; they coincided with the major land masses of the earth's surface, and they were convincingly isolated from one another by such obvious barriers as oceans and mountain chains. But were there only five regions? And were these five universally valid?

When observations concerning the other animal classes were added to Zimmermann's inventory of mammals, discrepancies among the regions began to appear. J. C. Fabricius had just proposed eight climates for the insect world, about half of them grouped around ocean basins rather than land masses. His seventh climate, for example, *occidentale*, included North America, Japan, and China.[52] Fabricius worked within a Linnaean framework, assuming that like climates would offer like forms. But even on the basis of the information then available, it was clear that this assumption was no longer tenable. In many places where temperatures were the same, insects—and animals in general—were known to be totally different. Contemporaries also found Fabricius's divisions vague and arbitrary and, if temperature were of primary importance, not sufficiently correct.

Pierre André Latreille later proposed an alternative division, which tallied better with the current state of knowledge, but this division, from the author's desire for precision and detail, obscured not only the primary divisions recognized by many entomologists but also the intelligent appreciation of distribution in all groups.[53] J. R. Forster published a handbook of natural history in 1788 in which birds, fishes, and insects occupied entirely different regions. Birds, for example, were given the entire Pacific as just one of their regions, while insects were divided among the American islands, New Holland, and the islands of the Pacific. The ornithologist, entomologist, and ichthyologist each had his own geography.[54]

Among naturalists whose insight was not obscured by the Linnaean *esprit de système*, Zimmermann's inventory of quadrupeds served as a model for the solution to the problems of distributing the other animal classes; the first prerequisite for any general statement was an accurate list of known species according to geographical location. When this was well and truly done, Illiger argued, the naturalist would not only know the habitations of all known animals, but would also be in a position, "by means of comparable surveys, of judging all organic productions in that part of the world and of concluding, on the basis of the one and the other and with reference to the mutual connection of these natural products in one country or climate, how one class determines another."[55]

Two of Illiger's works, his "Ueberblick der Säugethiere," based not only on Zimmermann but also on Bechstein, Erxleben, Humboldt, Meyers, and Pallas, and his "Tabellarische Uebersicht der Vertheilung der Vögel," based on Bechstein, Forster, Pallas, and Wolf (to name only German sources), are outstanding examples of this approach.[56] Illiger's tables cannot claim to be the origin of research into the principles of distribution of the larger systematic groups, but they certainly constitute a persuasive statement for the value of this kind of comparative study.

Another major problem for naturalists working with the idea of distributional regions was the presence of the same species in two or more regions. As we have seen, Zimmermann rejected the idea that distribution resulted from the migration of some species' members from the original zoological garden. He explained the presence of the same species in more than one region with the old analogy between form and climate. When, for example, "one discovered some specific kinds living in the torrid zones in both worlds, this proved not more and not less than this, that in both worlds in the same climate one pair of like animal kinds had been brought forth."[57] Zimmermann added immediately that these isolated facts did not alter the essentially endemic character of the fauna of a region.

Zimmermann's inventory had established to his satisfaction that by far the greatest number of animal kinds were limited by their natures to single regions and that they were further restricted by climatic barriers. And yet there were puzzling cases (in fact, the numbers were growing) of identical, apparently limited species whose populations appeared in two widely separated regions. Among naturalists who accepted the idea of regions of distribution, there was a strong tendency to question whether members of these disjunct populations could possibly be of the same species. Not only Zimmermann, but also Illiger, Rudolphi, and Willdenow, argued that even very strong resemblances undoubtedly concealed essential differences and that not even comparable ways of life could serve as infallible

Two Siberian passerines, Sturnus (Heterornis) dauricus *and* Alauda (Melanocorypha) mongolica, *described by Peter Simon Pallas in the Stockholm* Handlingar *of 1778. Pallas became a foreign member of the academy about this time.*

guides. Here, as in the enumeration of distributional regions, the dimensions of the problems could be indicated only by means of careful systematic work.

If these disjunct species were truly the same species, there could be no satisfactory explanation or even any general rule. When a naturalist found the same mammal or the same insect in two regions of the world, he was no longer justified in deriving that animal from a single one of those regions. When, for example, the steenbok appeared in both the Swiss Alps and some high ranges of Asia, on what basis was the animal to be derived from Asia alone? What was gained by making the assumption? "Even in a single part of the world," argued K. A. Rudolphi, "many parts of the same species might very easily have been created here and there."[58] Nothing spoke against this possibility. And the same power that had caused hyenas to appear in both Asia and North Africa "could also call them forth in Syria and Persia and so forth, and only in this way did they become useful early on in all these regions" (ibid., pp. 143–44).

The principle underlying this particular distributional pattern was incomprehensible to human understanding and would remain so no matter what conceptual schema was applied to the well-established facts. H. Fischer had written in 1771:

> Since we cannot assert that the world has been here from eternity or that the animals have crept forth out of the earth by themselves . . . so there remains only this, that all animals through the omnipotent hand of the Creator, each according to its kind, in a climate suitable to that kind, have appeared simultaneously everywhere on earth.
>
> . . . When animals of a single kind are found in different parts of the earth, and could in no way have been transplanted from one to the other, then more than one pair of these species must have been created, some in the Old, some in the New World.[59]

It is tempting to use the idea of regions of distribution, or centers of creation as they were later called, to stand for the whole of late-eighteenth-century geographical history. In the context of the period, though, the concept cannot pretend to such central importance. Most naturalists would not have identified the idea as a central theoretical issue. Orthodox systematists of all persuasions continued to ignore the problem of distribution until well into the next century, and even among those who took an interest in geographical questions, the number concerned with enumerating and coordinating distributional regions was small, necessarily limited to the sedentary few who, like Zimmermann, had access to large collections and libraries, who had a bent toward speculation, and whose duties allowed purely theoretical study.

The majority of working naturalists did not find it inconvenient to leave open the question of regional coordination. The materials and tactics that would eventually lead to new general ideas were, thanks to Zimmermann, far more easily recognized. The integrity of the period is best measured not against large-scale conceptual projects but against contributions by individual naturalists scattered throughout the proceedings of universities and academies. Among the most important of these were the many studies of Russian fauna published by Peter Simon Pallas between 1768 and 1811. Pallas was the first to describe carefully the range and mode of life of the Dseren antelope, the dschiggetai, and the extinct tarpan, the first to point out the singular fauna around Amur, and the first to excavate the carcass of a woolly rhinoceros—all sensations among the scientifically curious of that time.

One of Pallas's persistent concerns was the fauna of the inner seas and lakes of Russia—the presence of seals at Lake Baikal, for example, an animal that otherwise "is never in the habit of straying so far from the ocean in rivers, and in our time has not been observed in the Jenissei and the lower Angara." The fact was remarkable and could not be explained except by "an important alteration in the surface of the earth or by extraordinary and rare circumstances."[60]

Pallas later argued that the marine animals of the Caspian Sea had appeared there at a period when the Caspian and the Black Sea had been joined. "The seals, some fish, and marine shells that the Caspian Sea shares in common with the Black Sea make this former communication almost indubitable, and these same circumstances also prove that Lake Aral must have been joined to the Caspian Sea."[61] Pallas supported the argument with evidence drawn from the geology and geography of the region. When at the end of his long career he returned to the "very remarkable" presence of *Phoca canina* in Baikal, Oral, and Aral lakes, as well as in the Caspian Sea, he noted that they had reached there "imo diluvii Noachici tempore."[62]

Pallas also attempted to list carefully the then-known mammals common to the Old World and the New, and he mentioned the possibility of an earlier land bridge over the Aleutians, an idea also discussed by Fischer, who argued that the migration of both birds and land animals to the New World was impossible on the basis of present conditions. He discussed briefly the possibility of revolutions on the earth's surface and multiple floods, "so that what are now separated from one another by great bodies of water must formerly have been fast land."[63] But the idea was so encumbered with difficulties that Fischer felt obliged to assert that each animal had come forth from the Creator's hand, "each according to its kind, in a climate appropriate to that kind." Like Gmelin two decades earlier,

Fischer found no difficulty in the fact that God had placed the same species in two widely separated regions.

Another characteristic and important set of contributions to the geographical history of animals in the late eighteenth century is work published by the Forsters after their voyage with Cook.[64] A representative example of the older Forster's work in zoology can be found in a paper on penguins, "Historia aptenodytae," 1781.[65] The body of this work is an orthodox systematic description of a genus mentioned by every navigator who had visited the southern extremities of the globe, species of which had been described in detail by Anson, Byron, Bougainville, and Pernetty. At the time of the voyage, about ten species were known, and J. R. Forster proposed to define them "per genus et differentiam specificam." A conflict with Linnaean dogma is apparent in the subtitle of the paper, "Generis avium orbi australi proprii." According to Linnaeus, there were no genera proper to a single region only; the same climate everywhere was supposed to proffer the same forms. Although Forster nowhere contradicted Linnaeus, his copious detail on the habitations of penguins left no doubt that the creatures were indeed limited to some coasts and islands in the southern extremities of the globe.

After Zimmermann's fundamental work, the geographical history of animals continued for some years to consist in the discovery of new facts and new patterns of distribution. When, shortly after the turn of the century, attempts were again made to articulate this body of material theoretically, Zimmermann's method was still dominant in the German-speaking world. In the *Biologie*, Treviranus built on Zimmermann's technique of listing animal species peculiar to one region or common to several regions.[66]

Illiger conceived his studies of mammals and birds as enlarging and extending, on its own principles, parts of Zimmermann's work originally defective or wanting. Birds had not previously been surveyed according to geographical location, whereas the number of known mammals had doubled in the thirty years since Zimmermann's inventory. Not only had new kinds been discovered, a new part of the world first opened by Cook had become known; the natural classification had made extraordinary strides. All of this seemed to Illiger to justify new compendia, but in the manner of Zimmermann.[67] Rudolphi also based his essay on the distribution of the lower animals upon Zimmermann's "excellent work." He deemed Zimmermann's views alone appropriate to the worth and independence of the created world.[68]

In all of these cases, the treatment of animal geography was purely synchronic and simply accounted for the present state of distribution. The atemporal aspect of animal geography differentiated these attempts at syn-

Johann Reinhold Forster's illustration of Aptenodytes magellanica, *1781, an accession from the voyage with Captain Cook*

thesis not only from Zimmermann's but also from the parallel study of plants. Zimmermann's insistence upon an accurate picture of present habitations of animals was separate from his historical theses. Animal geography was not necessarily more historical than the classification of types.

The concept central to much of this work, the idea of multiple centers of creation, was what Candolle once called "un être du raison," and eventually it included most of the quandaries concealed in the problem of distribution. At times a center of creation meant only a point of origin for individual species. At other times it meant special centers where creative activity was especially intense—that fortunate spot, as Georg Forster put it, "that encompasses the entire stock of nature, the reserve of each climate and element."[69] Some naturalists linked the doctrine to the idea of historical development; others, as Hofsten has noted, "asserted that the present state of distribution had come into being through migration from the centers of creation and silently passed over the assumption that each kind had appeared in only one place."[70]

Geographical history, as it came to be understood by eighteenth-century naturalists, was in its earliest phase almost wholly a product of the ambiguous attitude toward spatiality within orthodox tradition. If, as most naturalists continued to believe, there were only as many natural forms as had been created in the beginning, then it was obvious that the location of these forms in space and time was of only secondary importance; whatever the influence of their location, natural forms remained essentially the same. Different spatial locations might alter specific forms to produce superficial variations, but "true" differences issued only from the hand of God.

It was equally obvious, however, that the distribution of manifold forms over the surface of the earth was not accidental. Organized bodies were adjusted to a particular way of life, so that each plant and animal seemed destined for the place where it was found. This information was considered indispensable for purely practical, utilitarian reasons and found its way into systematic work in the form of notes on physical locations (stations) and regional distribution (habitations).

When Candolle wrote what remains perhaps the best summary of older geographical history, he attributed all that was important in the discipline to the immense multitude of notes on the physical locations and geographical distributions of species in the accounts of travelers, in the collections of naturalists, in floras, in faunas, and in the great systematic works of the late eighteenth century.[71] I can only confirm Candolle's assessment. Geographical history was built with increasing confidence and success upon the separate but related studies of physical conditions, stations, and habitations. By 1800, German naturalists had created a generally accepted framework for the discipline.

The abstract principles that these same naturalists claimed to have deduced from the same multitude of notes cannot pretend to equal solidity. Given the fixity of natural forms and the recorded facts of distribution, it was natural to assume that design had preceded the deployment of form, that distribution was subject to a plan. But did the plan comprise a regulated unfolding from a single center or from a number of centers? In the confusion of opposing claims and justifications, the only secure foothold I have found is a procedural rule stated by Zimmermann, who said that the basis for any discussion of distribution was "the present cohabitation and subsequent partition within a country in and of itself on the basis of the present nature of the animals."[72] Clearly, Zimmermann hoped to turn discussion toward what could be established without argument—a salient feature, as I have tried to show, of some of the best work on geographical history between Haller and Humboldt.

Here, as in systematics, naturalists organized the facts better than they explained their assumptions and principles. An understanding of this situation is particularly important for the evaluation of the historical explanations prominent in the work of Zimmermann and Willdenow. A reflective reader cannot help being surprised at the ease with which younger naturalists—Illiger and Treviranus, for example—concentrated upon purely systematic elements in the work of their predecessors and ignored their historical theses. Ideas concerning the historical development of nature were tenuously attached to the conduct and discipline of exact observation and did not take root in the natural history of the later eighteenth century.

V

The Mechanism of Formation

Evolutio denotes the unrolling of parts already existing in compact form. When Albrecht von Haller in 1744 used *evolutio* as the Latin equivalent for *développement* and *Entwickelung*, he captured exactly the essential character of the then-dominant theory of generation.[1] To interpret organic formation as *evolutio* was to identify the process with the gradual unrolling or unfolding of a germ existing in the maternal egg or spermatic worm. The manifestation of structure resulted from the distension of a form present in the germ; the totality of the initial form determined the totality of the final form. And what seemed to be the successive formation of individuals over time was in fact the manifestation of structures that had been created simultaneously. The theory of *evolutio*, like the constancy of species, was part of a larger conception, a metaphysical view of nature, in which all existing beings, as necessary elements of order, had been in the world since the beginning.

Since the time of Regnier de Graaff (1641–73) and Nicolaas Hartsoeker (1656–1725), anatomists and physicians had disputed the relative merits of the ovist and animalculist versions of *evolutio*. Alternatives to the theory did not fare well, although they continued to appear. In Halle, for example, Georg Ernst Stahl (1660–1734) argued that the soul not only initiated but directed organic development and possessed direct knowledge of the organs and their functions, even in the embryo. Advocates of spontaneous generation continued to argue that organic forms might arise from matter by chance. In Paris, René Réaumur proposed a conception of generation in which living forms were independent of one another and issued immediately from the hand of God.

As controversy accumulated year after year, the system of *evolutio* established itself as the exclusive, universal, and specific mode of generation for living beings, opposed to any and all mechanist and materialist explanations. The old system of two fluids, for example, advocated the successive formation of parts from relatively unorganized matter provided by the prolific fluids of male and female. Evolutionists countered that, by rejecting the idea of a preexisting germ, epigenesists were left with ingredients, principles, and parts that lacked organic character and that could not, therefore, account for the source of organization in living forms.

During the first half of the eighteenth century, the system of *evolutio* came to dominate all study of generation. In 1746, after concluding a prolonged analysis of the field, Buffon said that ovists and spermatists "at present divide physicians . . . all who have written recently on generation have adopted one or the other of these opinions."[2] But there are good reasons for believing that this triumph had more to do with institutional inertia than with genuine conviction.

Neither ovists nor spermatists could account for so ordinary a phenomenon as family resemblance. Ovists argued according to the internal logic of their systems that the female contained a number of male germs capable of developing only once and a number of female germs capable of developing "to infinity." Among spermatists the situation was the reverse. The facile invocation of infinity constituted another problem. It was a rational abstraction, implying a division of matter on such a scale that the entire problem would have to be regarded as an illusion. Certainly it was well beyond any possibility of scientific verification.

This study follows the main lines of research in the science of generation between 1745 and 1790 and shows how French and German specialists challenged, criticized, and ultimately rejected the system of *evolutio*. The story includes an account of the alternative models for organic formation proposed in these years. But as the impressive secondary literature has shown, the emergence of epigenetic systems after 1745 involved far more than the peripeties of theory.[3] The epigenetic program cut across preconceived ideas, prejudices, and routines and provoked profound generational and institutional conflicts. Not even these conflicts, however, completely account for the significance that contemporaries attributed to the program. Everyone involved understood that a discussion of the capacity of living matter to organize itself was a declaration of the rights of nature in the name of rational science. Like the investigations of the system of nature and the scale of diversity, the epigenetic program opposed the prejudice of immutability in the world of living forms and treated nature as an autonomous power, pursuing, by means of her own forces, the continuities of her own development.

When Buffon sat down to write his treatise on generation in 1746, he had studied the problems of reproduction and development for over a decade.[4]

He had analyzed theories since the time of Plato, trying to identify the point at which science had taken a wrong turn. He had followed discussion in the Paris academy having to do with the capacity of matter to organize itself. The most important of these discussions concerned Abraham Trembley's reports on regeneration in freshwater polyps, but other examples of regeneration had been discovered in crabs, starfish, earthworms, and some kinds of seaworms.[5] Buffon had exchanged ideas on this research with Pierre-Louis Maupertuis, whose *Vénus physique* appeared anonymously in 1745. During the writing of his treatise, Buffon and John Turberville Needham investigated the existence of seminal fluids in both male and female. And Buffon continued to follow Needham's research. Buffon's treatise, then, rested upon a historical foundation, took account of recent research, and included some observation.

But the most important element in Buffon's program had nothing to do with the patient accumulation of evidence. Briefly, and perhaps too simply, Buffon—and Maupertuis—set out to reestablish the science of generation on the study of physical causes and intelligible principles. The two men undertook this reform with the firm intention of putting an end to scientific traffic in final causes and divine intentions. "Does science lose anything," asked Maupertuis, "if we think that animals are only formed successively? Is there any difference for God between the time that we regard as the same and that which succeeds itself?"[6]

The answer was obvious: God had nothing to lose, and science had everything to gain. No doubt Maupertuis understood far better than his younger colleague that investigation at the level of physical occurrence would confront many phenomena whose causes were unknown. But both men recognized that observation and experience would yield the only kinds of certainty of which physical objects were capable—at best, probable truths. Their abstract program included some recognition of its own limits. In place of the dominant theory of *evolutio*, Maupertuis and Buffon revived the discredited system of two fluids.[7] Their version of epigenesis, which proposed a causal explanation of generation with an active force as its mechanism, was to influence theories of generation for the rest of the century.

The basis for both the *Vénus physique* and the *Histoire générale des animaux* was an impressive history of generational theory, marred only by the assumption that the system of *evolutio* was inherently absurd. Ovists and spermatists, said Maupertuis, attached themselves stubbornly to the system, in spite of all the phenomena with which it was not compatible, and abandoned all hope for a natural explanation of organic formation.[8] As for the epigenesists, either they had got conception right and formation wrong, or the reverse. William Harvey, for example, convinced himself—incorrectly, according to Maupertuis—that he had destroyed the system of two fluids. But Harvey's observations of the process of formation were the best of any. Réné Descartes, on the other hand, had correctly championed the mixture of seminal fluids as the source of organic formation. But he had gone on to argue that purely mechanical causes formed body parts, views accepted by no one. In spite of these faults Descartes's project was preferable to that of *evolutio* because it returned the science of generation to the study of efficient causes and observable effects.

In the system of two fluids the problem was to find an intelligible mechanism to operate the formation of an embryo from the mixture of male and female fluids. The ancients, Maupertuis remarked acidly, experienced no difficulties on this score; everything was operated by a generational faculty (ibid., p. 81). It was clear that no one had yet explained how an embryo formed from the mixture of two fluids. And yet, Maupertuis added, "I believe it none the less clearly" (p. 119).

Maupertuis had followed the implementation of Newtonian attraction in other fields of scientific inquiry. He quoted his colleague M. Geoffroy, who had adapted attraction to chemical affinities: "Whenever two substances have some disposition to join one another and find themselves united together—if a third follows that has more relation with one of the two, it will unite to that by causing it to release the other" (ibid., pp. 120–21). Since Maupertuis did not draw a sharp line between living and inert matter, it was inevitable that he should apply this version of attraction to vital phenomena. "Why," he asked, "if this force exists in nature, would it not take place in the formation of the bodies of animals?" (p. 121). Affinity promised precisely the intelligible mechanism Maupertuis sought to operate the process of organic formation. "That there are in each of the semens parts destined to form the heart, the head, the entrails, the arms, the legs, and that these have a greater tendency to union with that which in the formation of the animal must be its neighbor than with any other; the foetus is formed: and if it were a thousand times more organized than it is, it would form itself" (ibid.)

This proposal seemed to conceive the process of formation wholly in physical terms, subject, like all physical phenomena, to other physical conditions. His model promised an intelligible account of family likenesses, as well as anomalous formations, to be explained only on the basis of contributions by male and female—or the successive apposition of parts.

Réaumur countered that attraction could scarcely operate the union of similar parts in a solid mass; it was hardly fit to account for the structure of a heart, a stomach, or an eye, much less their organization in proper order.[9] Maupertuis gave every appearance of conceding the point. "Uniform and blind attraction," he later admitted, "diffused in all the particles of matter, would not be able to explain how these particles arrange themselves to form the organization of the simplest body."[10] He turned to Leibniz, who had argued that the basis for any explanation of natural processes lay in simple substances whose nature could be described in terms of intelligence and appetite. As Maupertuis worked out this idea, he enlarged the notion of matter to include "something resembling what we call desire, aversion, memory." The material particles that constituted the living organism "remembered" their old situations and possessed an "instinct" to unite and arrange themselves in the embryo.

Buffon shared the assumption that organic formation consisted of the assembly of material particles operated by an external force. Unlike Maupertuis, Buffon drew a sharp line between brute and organic matter, and this complicated his version of the formation mechanism. Nourishment, development, and reproduction, Buffon argued, were effects of one and the same cause.[11] During the life of an organism, organic particles were taken up in food and separated from brute particles by digestion. The organic particles went to develop body parts with which they had some analogy. At maturity, the organism took in more particles than it needed; the surpluses were sent from all parts of the body to reservoirs, where the "extracts" came together to form a new organism. In the case of sexual reproduction, the new organism required the mixture of male and female seminal fluids to develop further. Buffon argued for the formation of small organic bodies in both male and female. These were, however, mere sketches "in which only the essential parts are formed."[12] Until the male fluid encountered the female fluid, these bodies did not develop.

The process of formation involved two components—the internal mold of the organism and penetrating forces—whose action was analogous to attraction in brute matter but operated only in organic matter. The internal molds were Buffon's alternative to preexisting germs. "In the same way that we can make molds to give the exterior of bodies whatever figure we please, let us suppose that Nature can make molds by which she gives not just the external figure but the internal form as well."[13] The animal body was just such a mold, and each part of the animal body was similarly an internal mold, receiving and shaping the particles with which it had most affinity and sending superfluous particles, imprinted with the mold of that body part, to the seminal fluid.

In the processes of nutrition and reproduction, the internal mold was, as the name suggests, inert and passive. To explain how the mold shaped organic particles, Buffon needed an agent, the penetrating forces "relative to the interior of matter and without any relation to the exterior qualities of matter, which act on the most intimate parts and penetrate them at all points" (ibid., p. 247). Organic particles, moved by these forces, flowed into the internal molds, augmenting mass and volume as the mold expanded. Simultaneously, the particles became identical with the matter of the mold. As for reproduction, a force similar to that necessary to produce growth operated the reunion of surplus particles, organizing them in a form similar to the body from which they came. The penetrating forces insured that the extracts from the body parts, with the internal molds they carried, were arranged in a new embryo.

Buffon convinced himself that this mechanism of molds and forces could account for the formation of parts that did not yet exist. In fact, though, he was describing the appearance of parts that were not only formed but identical with their mature state. By adding penetrating forces that were analogous to attraction but operative only in organic matter, Buffon rendered the process of organic formation wholly unintelligible.

Since both Maupertuis and Buffon conceived organic formation as the apposition of material particles, they required an external agent to organize those particles.[14] Their addition of force to matter and motion promised an intelligible physical mechanism, and this model continued to play an important role in the science of generation. But the nature of the force remained obscure. Simple attraction in and of itself did not explain the successive and unerring achievement of organic structure. Affinity in combination with representative particles from the parents' bodies was scarcely better. The basis for affinity, the similarities of the particles, was as obscure as the attraction they felt.

Maupertuis and Buffon championed observation and experience as the best sources of evidence, but their works were notably lacking in both. Let me mention a simple but critical case. The role of semen in the system of two fluids was obvious. Was there a corresponding prolific fluid in females? "I have investigated many times," Maupertuis wrote, "whether there are not similar animalcules in the fluid secreted by females. I have not seen any at all. But I should not wish to say that none exists."[15] He did not wish to say so because he could not, and so he added that females secreted many fluids, that confusion was possible, and that in any case the female contribution functioned differently from the male. Buffon hit upon the *corps glanduleux* as the source of the female fluid, but Haller demolished this observation. Neither Maupertuis nor Buffon troubled himself much over these contretemps. Buffon admitted that he was not a good observer. Maupertuis feigned indifference.

To explain the influence of the new system of generation, then, we are obliged to return to the point where we began, to the abstract position on the nature of science and scientific explanation. Maupertuis and Buffon revived the old system of two fluids because it promised to establish a firm connection with causes and processes proper to physical nature. Their attempt to conceive generation as a self-contained and self-explanatory process proved far more important than the success or failure of individual theoretical components. Observation, they implied, could give itself over to the study of physical causes and principles with confidence. The science of generation not only implied, but required, the study of other physical factors—heredity and environment, for example. Thus, while the epigenetic program lacked a sound observational basis, solved nothing directly, and involved intractable difficulties, it did promise to open up vast new areas to scientific investigation.

Between 1745 and 1755 a flood of pamphlets, memoires, treatises, and observations contributing to the new science flooded the learned world of western and central Europe.[16] Those members of the theological, philosophical, and scientific establishments who conceived nature as immediately and continuously dependent upon God reacted, as intended, with consternation and hostility. Their automatic response was to defend an entrenched mind-set. The guilty parties responded with public concessions. Buffon informed the Sorbonne that his theory of the earth was a "purely philosophical supposition" that did not question revealed truth.[17] When, however, he was attacked in *Nouvelles ecclésiastiques*, he wrote a friend, "Chacun a sa delicatesse d'amour-propre, la mienne va jusqu'à croire que de certaines gens ne peuvent pas même m'offenser" (ibid., p. 702).

The German translation of Buffon's treatise on generation brought Albrecht von Haller face to face with this suspect product of French materialism. His preface to the translation also appeared in French in 1751 as a separate publication.[18] Haller's tone was courteous, benevolent, and magisterial. Because he so openly admitted his opposition to Buffon, some commentators have taken his claims to impartial truth seriously. We should beware of doing likewise. Haller's intention throughout his preface was to return nature to the divine jurisdiction and science to the contemplation of His works. Toward this end, Haller did not shrink from misrepresentation and distortion.

Here is a simple example. Buffon's first chapter concluded with the provocative phrase, "Life, far from being a metaphysical degree of being, is a physical property of matter," a phrase that so enchanted Diderot that he took it as his own.[19] Catholic apologists were fond of it, too, but as a subject for sarcastic gibes. Haller approached the matter indirectly. Pursuing the continuity of organization through the three kingdoms of nature, he, like Buffon, drew a sharp line under *plants and animals* and concluded that "life is a higher degree than mere vegetation or crystalization." A higher degree of what? One can only conclude, a higher degree of being. "No one," Haller went on complacently, "will accuse me of partiality in the exposition of doctrine."[20] The situation is familiar: one expert sums up the testimony of another and, in the process, destroys it. Haller's exposition of Buffon was full of similar impartial truths.

Haller's preface first explained why, for purely scientific and technical reasons, he could not accept the views of "that ingenious Frenchman" (ibid., p. 3). A second part explained why the materialism of Buffon and Needham was not a threat to truth (pp. 3–4). The arrangement betrays a deeper intention, however. Haller, focusing discussion on the technical matters in which he was an indisputable master, would expose Buffon's scientific inadequacies. Haller could then devote the rest of his preface to the implications of Buffon's views, confident that his own arguments would accommodate what few truths Buffon and Needham had uncovered.

Haller's phenomenal erudition was more than a match for Buffon's. The system of two fluids was an adaptation of "the opinion of the ancients that lasted up to the time of Steno" (ibid., p. 26). The idea that seminal fluid flowed from all parts of the body revived the Hippocratic doctrine of *defluxus*, "an idea differing widely from accepted opinion among evolutionists" (p. 17). As for the spermatic animals of Ham and Hartsoeker, Haller inclined more to the opinion of Needham, who saw them as true animals, than to Buffon's view, "that they were organized particles" (p. 29).

Haller conceded that the system of two fluids accounted for family likenesses and a "number of questions almost without solution in the doctrine of *evolutio*" (ibid., p. 27). But every concession was accompanied by an even more profound objection. Female seminal fluid, the basis of Buffon's theory, did not exist. "Who has seen it, and who has ever found in the body of a woman anything resembling the seminal fluid of the male?" (p. 49). The *corps glanduleux* could not possibly serve as a cause of fertilization; they appeared only after conception (p. 47). Buffon had not only committed a fault against science, he was not a particularly good guide in matters of feminine anatomy. "Buffon's system falls, without hope of resurrection" (p. 48).

Family likenesses were not a scientific subject, said Haller. Anatomy showed that there had never been two men whose nerves, arteries, veins, muscles, and bones were not entirely different. Nature was not simply far from uniformity; her variety was so great that we might believe she worked without a plan. But if a foetus was only a copy of the parent, why did foetal structures contain so many parts not found in the parents? If children were formed of extracts from parts of the parents' bodies, how could deformed parents produce normal children? (ibid., pp. 33–38).

Buffon's account of the mechanism of formation was the weakest part of his theory. The internal mold revived *evolutio* in its most primitive form. "The general model of humankind," Haller wrote, "issues immediately from the hands of God . . . Nature . . . only copies molds already created" (ibid., p. 66). Haller turned the arguments concerning scale and liquidity, which Maupertuis and Buffon had used with such devastating effect, against Buffon. "Is it possible that from tenacious matter nature could produce a being infinitely small, perfectly like the parent, and whose blood, for example, would surpass infinitely in delicacy that which flows in the veins of the father?" (pp. 30–31). The answer was so obvious Haller simply dropped the argument.

Turning to the penetrating forces, Haller constructed an argument remarkably like that of Réaumur. No force in nature could arrange so many parts in their proper order. Even if one granted the operation of an internal mold, one could not explain how so many parts in the seminal fluid came to be arranged in a foetus. Buffon "has need here of a force that has eyes, that makes a choice, that proposes a goal, that against the laws of blind combination always leads, and ineluctably, to the same body" (ibid., pp. 42–43). Haller was not rejecting the force of attraction; he was as much an attractionist as Maupertuis or Buffon. But not in quite the same way. The forces of matter were responsible for a chain "of organizations extending without an interruption from the organization of an Alexander to a snowflake" (p. 24).

But could the forces of matter alone produce an Alexander? Obviously not. "If matter has forces that allow it to build things, it does not possess them blindly. They are surrounded by eternal limits and always build perfectly, not mechanical equals but something similar, something that is prescribed in an inviolable plan but with a diversity that includes the constraints of blindly working matter" (ibid., p. 60). Matter had only those capacities and forces allotted to it and calculated according to a plan. Obviously, living beings were not formed by blind combinations of matter. Haller went further than Buffon in his admission of diversity in the process of formation, but he was also much stricter in his interpretation of regularity. "If the extensive and attractive forces of seminal fluid alone form the man or the deer, if it is only by chance that this occurred, why then does not this matter . . . produce in place of a man, an ape, who resembles man so much?"[21] The constancy of species included some variation but never deviation in essential characters.

Haller concluded, "The invariable reproduction of animals, always alike and always divinely well constructed, seems to be beyond simple forces that produce only a simple conserve, a salt, a crystal, or a microscopic animal, lacking a heart and members, and whose life consists simply of irritability, and whose form is uncertain" (ibid., p. xvi). It is no exaggeration to find in this passage the reaffirmation of a science of fixed forms and final causes. Haller did not insist upon a purely mechanical and passive nature; there was simply nothing to be feared from the autonomous activity of matter, since this was subject to eternal limits. "If matter has forces that make it able to form itself," he said, "it does not receive them from blind fate. They are enclosed in eternal limits; they always form perfectly, not beings prescribed in an inviolable plan but with a variety that excludes constraint from a blindly efficacious matter."[22]

It was not the mode of production or the course of development that proved the existence of God; it was rather the relation between structure and design. The nature of Buffon and Needham was therefore no more prejudicial to religion than that of Newton. Buffon's system was perhaps even less dangerous than that of Needham. Buffon's internal molds, the source of organization, could have come only from the hand of God. It was true that Needham denied the existence of anything preformed in the seminal fluid, but for just this reason he had been unable to account for the reproduction of animals always alike and always perfectly made (ibid., p. 54).

It is striking how many of Haller's most telling arguments were taken from the evolutionist armory. The partisans of *evolutio* had always assumed that any theory based on the apposition of parts was bound to fail. By abandoning the idea of a totality irreducible to mechanical composition, epigenesists also abandoned any possibility of accounting for the process of development. Haller simply adapted the argument to Buffon's complicated formation mechanism. The internal molds revived a particularly archaic form of preformation. As for the agent of formation, the penetrating forces, Haller argued that these forces were purposive agents. If the critical issue in any theory of generation was a viable source of organization, it is not unfair to say that Haller was already a member of the evolutionist camp in 1751. He did continue to equivocate: as late as 1754 he wrote Charles Bonnet, "I see more and more a simple, sticky material forming and shaping itself little by little."[23] But after Haller returned to his native Bern (in 1753), he undertook the observations of incubated chicken eggs over the course of three summers (1755, 1756, and 1757) and converted to the doctrine of *evolutio*.

Haller's concern, before he began his observations, was the initial fluid state of the embryo; he believed that a transparent fluid congealed to form foetal membranes. Bonnet suggested that the transparency concealed structures that could be revealed with spirits of wine. Haller eventually agreed. The small animal that became visible with the help of spirits of wine on the eighteenth day had existed the day before, approximately the same, "but its transparency hid it from us" (ibid., p. 58). Haller's argument from invisibility was to play a role in his theory of *evolutio* exactly analogous to the formative forces in epigenetic theories. Invisibility became Haller's all-purpose source of organization, to be invoked on all occasions—but with the major scientific flaw that it was unverifiable.

When Haller began his observations of incubated eggs, he concerned himself with the formation of bones, heart, and lungs. His guide in matters concerning the heart was Marcello Malpighi, who had stated that the heart began as a simple tube that later developed chambers. Haller could not reconcile the circulatory apparatus of the lungs with a tubular heart. If the heart curled up upon itself to form a four-chambered heart, the vessels of the lungs would break at some point in the linear arrangement. Haller reported to Bonnet that the heart "was always equally formed of two auricles and two ventricles" (ibid., p. 73). A second set of observations confirmed this opinion.

> For the heart, all the mystery that I investigated is reduced to a very simple thing. The lung, with its artery and vein, is very small and invisible in the first times. Little by little it grows, and the artery which through its fineness and its transparent liquid was hidden from our eyes becomes a second very considerable branch of the aorta, from which two small branches lead to the lungs. (ibid., p. 89)

The lungs and their vessels, complete but invisible, were connected to the heart as they were when they became visible. The heart, too, developed from four originally existing chambers, also invisible. The whole apparatus became visible during development, but no essential transformation took place.

Haller was ready to announce his conversion to the doctrine of *evolutio* by the summer of 1757. The decisive evidence had to do with the relation between the egg yolk and the embryo. In later stages of development, the intestines of the embryo were connected to the yolk sac in such a way that the yolk sac seemed to continue the intestine. Haller thought he had found the membranes of the yolk sac in unfertilized eggs. He concluded that the embryo must also exist at this time, although it was, quite predictably, invisible. As Bonnet explained, "the yolk sac is an expansion of the small intestine of the chicken. Thus the embryo existed in the female before the approach of the male" (ibid., p. 111).

Haller announced his conversion in the two volumes of *Sur la formation du coeur dans la poulet*. In every essential respect, an initial form in the egg determined the final form. Upon conception, the heart began to beat, fluids circulated through the folded parts of the transparent embryo, embryonic structures solidified, and a visible foetus emerged. As Haller explained,

> it appears very probable that the essential parts of the foetus exist formed at all times, not it is true in the way they appear in the adult animal; they are arranged in such a fashion that certain prepared causes, hastening the growth of some of these parts, impeding that of others, changing positions, rendering visible organs that were formerly diaphanous.[24]

Haller's observations solved many of the problems he had raised in his criticism of the system of two fluids. What seemed to be the successive formation of parts was in fact the gradual manifestation of parts. Unable to accept autonomous material forces or immaterial vital forces, Haller was left with simple mechanical forces. As Shirley Roe has remarked, *evolutio* was then the only viable solution to the problem of organic structure.[25] Haller's argument continued:

> If the first rudiment of the foetus is in the mother, if it has been built in the egg and has been completed to such a point that it needs only to receive nourishment to grow from this, the greatest difficulty in building this most artistic structure from brute matter is solved. In this hypothesis, the Creator Himself, for Whom nothing is difficult, has built this structure. He has arranged at one time, or at least before the male force approaches, the brute matter according to foreseen ends and according to a model preformed by His wisdom.[26]

Haller's treatise reaffirmed the essential tenets of the doctrine of *evolutio*. Organic generation consisted of a structure preexisting in the maternal egg. Development was in essence the distension of something that already existed and programmed to unfold. The totality of the initial form determined the totality of the terminal form—a living organism, separate from the parent.

Haller's research established the problem of organization as the central issue in the theory of generation. His argumentative superiority resulted from his domination of every aspect of the problem. With tremendous energy he heaped up the hitherto unobserved facts, adapting and inventing arguments in support of his position. Overwhelmed by his evidence, his authority, and his erudition, few can have noticed the frequency with which he invoked the initial invisibility of parts, and fewer still would have dared confront him openly.

A year after the publication of *Sur la formation du coeur dans le poulet*, Haller received from Berlin a dissertation on the theory of generation with an accompanying letter from Caspar Friedrich Wolff.[27] Haller's reaction can be surmised. Wolff's theory was a compendium of the errors Haller had just confronted and overcome with his conception of ovist *evolutio*. Not only did Wolff argue for successive formation from unorganized matter, he proposed a formation mechanism operated by an external force. All of this was perhaps excusable. Wolff admitted that his dissertation had been practically finished before the appearance of Haller's treatise. But the argument was supported by an armature of scholastic definitions, principles, laws, and scholia that offended Haller's sense of scientific discourse.

Thirty years after the dissertation, Wolff summarized his early work. He had not been concerned, he wrote, with the complete formation of any one part but only with the general processes of production and organization common to plants and animals.[28] His inquiry began with the appearance of new parts from an originally fluid state. Microscopic observation showed that a new part in a young plant began as extruded sap, which had neither the figure of the part it would become nor any inner structural organization. "It was impossible for me," he wrote, "to conceive how parts already brought forth could then become organic and how, when they had been formed without vessels and vesicles, could acquire both vessels and vesicles" (ibid., p. 49). After repeated observations, he discovered that every part of a plant or an animal was first produced, then organized. "When this had been discovered," he said, "everything else was easy." All vegetation—that is, all creation and modification of organic structure—began with the excretion or deposition of fluid. As a deposit was nourished by more fluid and grew, it solidified. Fluid continued to flow, pushing through the original deposit, forming vessels and vesicles as it passed. This, wrote Wolff, was the basis for the theory of generation.

As an example of the process, Wolff took the liver, which began as small vessels from which other vessels arose, resulting in the complete system of liver vessels. The coverings of these vessels exuded a fluid that formed the cellular tissue binding the vessels together. Finally, the whole mass acquired a covering, after which the liver increased in size but acquired no new organic parts (ibid., p. 48).

The circulation of fluids was central to Wolff's conception of formation. But since the heart, like other organs, formed only gradually, and since the movement of fluids took place before the heart began to beat, and since, finally, the circulatory mechanism included both plants and animals, Wolff needed an agent above and beyond the mechanical action of the heart to explain circulation. His *vis essentialis*, or essential force, at least in this early version, did not improve upon the attractive and penetrating forces of Maupertuis and Buffon.

The essential force (whose existence was to be inferred from observation) operated the extrusion and organization of fluids. Plants absorbed fluids from the soil and, at the tip of the growing stalk, produced leaves, flowers, and fruit in serial order. The essential force "must be assumed if plants and their nourishing liquids are granted, which has been confirmed by experience. This will be sufficient for the present purpose, and it will be called the essential force of plants."[29] Similarly, in incubated eggs, an embryo increased in size from the first moments of incubation, before it possessed a viable heart. "Material must be entering the embryo from the yolk, and it follows that the nourishing particles pass from the egg into the embryo and that there exists a force through which this is accomplished. . . . I call this the essential force" (ibid., p. 73).

With the extrusion and solidification of fluids and the essential force, Wolff believed he had conceived a complete mechanism for the general process of vegetation. "The essential force, along with the capacity of nutrient fluid to solidify, constitutes the sufficient reason of all vegetation in both plants and animals" (ibid., p. 115). In keeping with his philosophical method, Wolff recast the process of vegetation as a combinatory involving simple elements and their modes of successive composition. "One truly explains generation," he wrote, "who deduces the parts of the body and the mode of its composition from known principles and laws" (p. 5).

To a degree, Wolff forced his observations into the straitjacket of his deductive method. His vessels and vesicles were simple elements, building blocks, from which formation could be reconstructed rationally. Similarly, the essential force was an *être du raison*, not a product of observation. The problem of providing a causal explanation for the successive organization of organic forms led Wolff to the inner bond between nourishment and structure. He viewed the cooperation of these two processes as a force whose effects separated organisms from all other kinds of composition. It would be tempting to reject Wolff's theory of vegetation as a rational reconstruction, if his observations were not among the most acute in all of eighteenth-century science.

Let me consider a few of his observations, focusing on those that Haller found the most disturbing. In the later summary of his dissertation, Wolff said the formation of the heart was at the center of his differences with Haller.[30] In Haller's version of *evolutio*, the mechanical action of the heart not only initiated the circulation of fluids but was the chief agent in the emergence of a visible foetus; the heart thus necessarily existed at all stages of development of the foetus. Wolff's position was more complex. The heart, like any other organ, formed over time: in an embryo of twenty-eight hours, no heart could be seen; at thirty-six hours, the heart was a white ring, not connected to vessels and not beating. Before a heart existed, the essential force initiated the movement of fluids, supplemented much later by the beating of the new heart. Wolff returned to the plate by Malpighi, the very image Haller had rejected, to illustrate the newly formed heart. The heart first appeared as a looped tube, consisting of the left heart chamber, into which the end of a vein was inserted; at the other end an artery emerged.

An issue connected with the appearance of the heart was the formation of what came to be called the *area vasculosa*. For Haller, the *area vasculosa* was an example of preexisting membranes gradually unfolding to reveal their length and breadth. In Wolff's account, fluids moved by the essential force gradually broke up granular material in the *area*, forming paths without membranes and then becoming perfected small vessels.[31] Finally, Wolff attempted to establish the essential difference between plants and animals in the absence or presence of a heart. Because plant substance solidifies more quickly than animal substance, plants can form parallel vessels only as fluids pass through. Because solidification proceeds more slowly in animals, there is time to form branching vessels. The heart arises only in situations where smaller vessels are gathered into larger and these, finally, into one common vessel, which, whatever its form, is itself the heart (ibid., pp. 92–93).

Georges Canguilhem has pointed out that Wolff's concern in the dissertation was to offer an explanation of development in general as well as original observations on the formation of the chick. He was not, however, primarily concerned with the concept of formation that these implied.[32] Early reactions to the dissertation, recorded by Wolff himself, have to do with Wolff's imaginative feat, a slow-motion study of the growth of a chick. Take Büchner's reaction, for example. "I have always held to predelineation," he told Wolff. "But now one sees clearly not only that the parts are formed successively but also that this formation takes place by the mode of excretion."[33] It was Haller, seconded by Charles Bonnet, who insisted (precisely as they had done in the case of Maupertuis and Buffon) upon the dangerous novelties in Wolff's conception of formation.

Haller's review in the *Göttingsche Anzeigen* politely pulled Wolff's theory to pieces. The essential force, that putative source of vital activity, derived from Needham's vegetative force. In the formation of plants, Haller argued, Wolff's efforts turned upon showing "not that parts were small and transparent at first but that they did not exist and that the vessels and vesicles from which all else formed were the consequence, and not the cause, of the movement of fluid."[34] If true, Wolff required no more than the solidification of sap and a basic force of movement to explain organic formation.

As for the generation of animals, Haller took immediate exception to Wolff's principles "according to which what one cannot see does not exist." The reason for this is that in animals everything is constituted from small spheres, and these are visible; consequently, no part can be accepted that is invisible and yet may be present" (ibid., p. 138). The issue, as both men knew, was not a theorem peculiar to Wolff's dissertation. It involved the central issue in the science of generation, the source of organization attained by a developing embryo.

Anyone adept in the use of a magnifying glass, Haller continued, knew that strong colors made parts visible and that transparency made them invisible. The remark linked Haller's rejection of Wolff's general thesis with Wolff's treatment of the *area vasculosa*. Wolff had described roads being cut and then becoming proper vessels. "The appearance is correct, but the doubt remained whether the transparent roads among the granular matter are really only roads without membranes, and this is not so easy to decide" (ibid.). Wolff's account of the formation of vessels in the *area vasculosa* was an example of the reasoning that led him to argue (incomprehensibly, according to Haller) that the embranchment of vessels led to the formation of a heart (p. 139). And this, exclaimed Haller, was supposed to constitute the essential character of animals.

Haller concluded with Wolff's treatment of the heart. The heart supplemented the action of the essential force but only at a late stage of development. Not only did Wolff claim to have seen a heart that had not yet begun to beat; he asserted that the heart beat only when it transmitted red blood. "The whole of mechanical medicine," Haller claimed, "to the degree that it seeks to explain the movement of the body on the basis of structure, is a mere chimera" (ibid., p. 140).

This was the most destructive piece of criticism that Wolff ever received. The review repeatedly acknowledged the importance of Wolff's work, but Haller also implied that Wolff was a young man in a hurry, that he was drawing hasty conclusions, and that he threatened to destroy the whole of mechanical medicine. Among colleagues in Halle and Berlin, Haller's review may have confirmed their suspicions about the young doctor's soundness and ambition. Wolff's dissertation appeared without a faculty sponsor, an indication that his professors found him too original or too independent.[35] In any case, Wolff took service with the Prussian army at Breslau under the patronage of the royal physician, Cothenius. When Cothenius later requested the Berlin Collegium to permit Wolff to lecture on physiology, the members of the corporation replied that Wolff's visit was "nicht anders als nachtheilig." Cothenius then imposed his protégé on his former teachers, and Wolff gave his course. But he was passed over when the collegium filled two posts that year. Wolff was thirty-three years old when at last he accepted a post at the St. Petersburg academy.[36]

While Wolff was lecturing in Berlin, he prepared a sharply worded defense of his position, the *Theorie von der Generation in zwo Abhandlungen erklärt und bewiesen*. The science of generation, according to Wolff, not only differed from physiology and anatomy but took precedence over them. Moreover, generation was not just observation and experience; it involved a philosophical component. On the basis of known principles and rational causes, the science of generation "explained" why an organic structure must be thus and not otherwise. Only an epigenetic theory could satisfy this philosophical requirement, because only epigenesis derived the phenomena of generation from physical causes.[37]

Wolff offered a history of various theories of generation, as had Maupertuis and Buffon. Wolff's was more perfunctory than theirs, scarcely more than a *catalogue raisonnée* of mistakes. He concluded with a refutation of *evolutio*, or as he called it, *predelineation*. Nature offered nothing that resembled a development that did not also include production—with the possible exception of metamorphosis in insects—or even comparable cases from which analogies could be drawn (ibid., pp. 14–31).

Wolff then turned to the defense of his version of epigenesis. His differences with Haller were differences of observation, not principle. Haller claimed that Wolff had built upon a hasty equation between visibility and existence. But Wolff had derived his statement "what one does not see does not exist" from observation. He went over the formation of vessels in the *area vasculosa* and asked, "Do I assert that there are no vessels because I do not see them and make this conclusion on the principle that what I do not see is not there? [Or] rather, do I not conclude these truths from that which I really see, because I see that . . . the vessels are not finished, but . . . are scarcely . . . begun and . . . not yet vessels?" (ibid., p. 88).

As for Bonnet, Wolff later admitted to Haller that he had been too harsh.[38] Bonnet had drawn far-reaching conclusions from Haller's observations of the continuity of membranes between the hen and the embryo. "I do not see the connection between these assertions," Wolff wrote. "One part of a thing continues directly into another part of the same thing; hence the one part could never have existed without the other, and the whole thing must have been produced *in instanti*. Only Bonnet has concluded thus, not von Haller."[39] Wolff then reprinted Haller's review of the dissertation without Haller's permission, an impertinence for which Haller reprimanded him.

The second part of *Theorie* was very different from the aggressive defense of epigenesis in the first part. This short treatise is the most readable exposition of Wolff's conception of formation. Again, the exposition took the form of a rational reconstruction. He first resolved organic parts into simple elements. Then he classed them according to mode of formation and relative autonomy. The developmental mechanism consisted of the deposition and solidification of nutritional fluids plus the essential force (ibid., 163–64). The distribution of fluids was not by means of a hydraulic pump, but "the determination and ramification of fluids depends solely upon the division of fluids and has a very different cause as its basis." This cause was, of course, the essential force, "that force through which in the vegetative body all those things are accomplished on whose basis we attribute life to it." The essential force was, however, a regulative principle, not a constitutive agent, and it was not the crux of the theory. "It is enough for us to know that it is there, and . . . we know it from its effects, as [it] is purely and simply required to explain the formation of parts therefrom" (ibid., p. 160).

In explaining the formation of vessels and the cellular tissue of animals, Wolff restated his argument, that the presence of a heart was the essential difference between plants and animals. Plant substance differed from animal substance because, in solidifying rapidly, plant vessels did not grow very large. If a leaf, for example, required more sap, new parallel vessels would grow in the stem, and just as many vessels would grow in the leaf as there were vessels in the stem. In animals, whose substance solidified slowly, vessels branched off from previous vessels. When a number of vessels formed in any one part, they were all branches of one vessel. Their embranchment implied a "necessary terminus," a simple trunk from which all arteries arose and to which all veins returned. This vessel was the heart.

Like all other vessels, the heart originated as a branch of another vessel—a vessel belonging to the mother, not the embryo. As other vessels branched from the heart, the heart changed. Fluids previously drawn from the maternal vessel were now taken from the substance of the egg. The first branches of the veins loosened a substance in which the heart cavity first took form. The heart then pulled away from the maternal parts, having found a new source of fluids.

> The heart, which was previously a branch of the maternal vessels, and which drew fluids from these vessels, now ceases to do so and takes them from the veins, which, like the arteries on the other side, are formed simultaneously; and the heart becomes what it had not been otherwise, the single independent source of all arteries, and the confluence of all veins. (ibid., p. 186)

In an appendix called "repeated experiments," Wolff returned to the main points in his developing controversy with Haller: the formation of vessels in the *area vasculosa*, the action of the heart, and the continuity of membranes between hen and embryo (ibid., 257–58). Observation of the earliest stages of the *area vasculosa* showed that the *area* could not have been formed by the action of the heart. Haller's vessels were in fact cavities in the granular substance. These did not communicate: they were so enclosed that the fluid they contained did not run out. The cavities were as wide as they were long, not at all like vessels, and were more numerous on the periphery of the *area* than closer to the embryo. They were not vessels or even paths cut by liquid pressed from the heart. "These cavities do not communicate with one another, and no passages toward the heart are present through which the liquid would have had to come" (p. 262). In an accompanying letter to Haller, Wolff insisted upon the proof of progressive formation. Because the islands and cavities first formed on the periphery of the *area* rather than near the embryo, he argued, they could not have been formed by the action of the heart.[40]

As for the formation of the heart, Wolff argued that it did not exist before the animal was twenty-four hours old and did not beat until much later. He had observed a heart at twenty-four hours and was astonished to see a feeble movement, a soft, gentle compression of the blood in the heart chamber, which remained unmoved. "Movement," said Wolff, "does not arise all at once, as I once imagined, but by degrees. It gradually becomes stronger and at last finally passes over from a slow spastic movement, as I have described it, into an instantaneous convulsive movement."[41] In other words, Wolff claimed to have observed not only the gradual formation of the heart but the development of a heartbeat. "Is not this sufficient," he asked, "to prove my theory of the essential force, of the gradual formation of parts and the gradual attainment of qualities characteristic of the adult form? I shall leave the whole matter to the judgment of Herr von Haller" (ibid., p. 270).

One of Haller's decisive proofs of *evolutio* was the continuity of membranes between the yolk and the embryo's intestines. On this basis, he declared that a hen possessed the true beginnings of the foetus before copulation.[42] Wolff argued, however, that the continuity was faulty observation: the membranes of the yolk were not identical with the original membrane covering the unfertilized egg. As development proceeded, the embryo formed inside the original membrane; it was connected at the intestines with the yolk sac, just like the *serosa* covering the embryo; but these were formed during incubation, later than the embryo itself. Haller's hastily construed continuity fell apart and, with it, one of his proofs for *evolutio*. Even if, Wolff continued remorselessly, continuity of the kind described by Haller existed, mere coexistence did not prove simultaneous creation, as Bonnet had claimed.

Wolff offered these criticisms with some deference, but all of them pointed to a single conclusion: that the succession of phenomena under the microscope could not be reduced to a single structure that maintained its identity throughout formation. Haller never had another opponent who challenged him so thoroughly on his own ground—his mastery of close observation and technical detail. Wolff could not match Haller's erudition, but he got the better of Haller as a tenacious observer and a relentless critic (ibid., p. 60).

In the end, though, Haller remained faithful to his original thesis. All of the phenomena discussed by Wolff could be seen as something other than progressive formation. Haller was most troubled by the formation of vessels in the *area vasculosa* (ibid.). By August 1765 Haller had convinced himself that the vessels preexisted but were very fine. As for the heart, "which this physician claims to have seen born," Haller never wavered. He could not afford to do so. The preexistence of the heart was central to his theory of *evolutio*. Wolff's remarks on the continuity of membranes between the hen and the embryo forced Haller to review his account, but he conceded as little as possible. The invisible embryo was continuous with the yolk, and the yolk sac membrane came from the mother.

Wolff was very much on Haller's mind when he wrote his summa of generational theory, volume 8 of the *Elementa physiologiae*. Haller marshaled his evidence one last time, refuting his opponent point by point. He saw the *area vasculosa* "as does the illustrious Wolff: we differ in our conclusions . . . I believe I can take for granted that the vessels have had membranes at all times" (ibid., p. 63). He had seen the heart "fully formed and perfect at forty-eight hours. If it was not visible earlier, it was because of its smallness, its transparency, its whiteness, and its resemblance to a mucous cellular substance concealed in it" (p. 69). As for the continuity of membranes between the hen and the embryo, the intestines of the embryo were identical with the original membrane (p. 78). In all of these cases, development was a matter of manifestation—by growth, solidification, opacification, and coloration—of something at first invisible. Changes in embryonic forms, once visible, were changes in the relative situation of parts or were due to different rates of growth.

Haller reserved his heavy artillery for Wolff's formation mechanism. The argument had not changed since Haller first used it against Buffon. Why, Haller asked, does the essential force, which is unique, form parts in the animal that are so different, always in the same place and always according to the same model, if inorganic matter is changeable and capable of assuming all sorts of forms?

> Why does this force always build without any error a chicken out of the unorganized material of a hen, a peacock out of a peacock? Nothing is assumed other than an expanding and progressive force. I would expect from this only that the net of vessels would necessarily become larger, as long as the expansive force can overcome the resistance. Why, at the site of this net, are formed a heart, a hand, a brain, a kidney? Why in each

> animal is there a proper arrangement of parts? To these questions no response is given. (p. 113)

In fact, no response was possible. The forces we find in the works of Maupertuis, Buffon, and Wolff are the Achilles heel of their version of epigenesis. Haller could not have known that Wolff was already trying to move beyond this particular set of assumptions. Wolff's early work centered upon the vegetative function common to plants and animals. He had shown in a general way how parts formed in serial order, starting with the production and organization of vessels and vesicles. He had not, however, attempted the description of the complete formation of any single part. In *De formatione intestinorum*, he undertook a complete representation of the formation of the intestinal canal.[43]

Wolff intended his analysis as a proof of epigenesis. If invisibility did not mean nonexistence, as Haller continued to argue, Wolff had to prove nonexistence. One way to do so was to establish the incompatibility of the invisible part with existing realities. Wolff did so in two different ways. He argued that there are modes of formation not to be reduced to any scheme of *evolutio*, processes such as the folding of tissue, the closure of an open organ, and the fusion of two sheets in a tube. Wolff also argued that formation often operated by substituting one part for another; *B* could not have been present from the beginning, since *A* occupied its place. The theory of *evolutio*, however complex, was no more prepared to admit destruction than formation.

By way of introduction, Wolff contrasted the generation of plants with animals. In plants, the same cause was everywhere at work, and the result was a simple, systematic product in which some parts furnished the basis for the existence of others (ibid., pt. 1, §9). In animals, only the extremities could be described as products of simple vegetation, analogous to the formation of plants (§§11, 12). Everywhere else, multiple modes of formation lay at the basis of multiple modes of activity. Among them an observer could not discover any relation, at least not such that one furnished a basis for another (§9). The different products—the nervous system, the circulatory system, the pulmonary system, the intestinal canal, and so on—were united in a single body. For these reasons, it was difficult to state the principles of generation and modes of formation for animals.

The formation of the intestinal canal offered a prime example of these difficulties. Wolff's treatise developed out of his earlier work on the *area vasculosa*, beginning with three systems of veins and their embranchments (§§17–20). Throughout the treatise, Wolff patiently charted the location of veins and arteries, but circulation was for once a subordinate consideration. Wolff was far more interested in the formation and dispersion of tissue, particularly during the earliest hours of formation.

At the center of the *area* he observed a transparent region formed when the upper and lower membranes of the *area* parted to accommodate the embryo (§§22–23). Harvey, Malpighi, and Haller had described the region and called it, variously, the *chorion*, the *amnion*, and the *yolk membrane*. Wolff saw it as a temporary enclosure for the embryo, an enclosure destined to disappear after the embryo was enveloped in more enduring membranes, the false and the true amnion (§24). He devoted the rest of the first treatise to the false amnion.

At the beginning of the second day of incubation, a part of the lower membrane of the *area vasculosa* swelled, folded sharply, and formed a pocket, first in the head region, then in the tail region of the embryo (§36). Wolff called the pocket the *false amnion* and compared it to the pupa of a butterfly (§37). As in the case of the earlier transparent region, the sac of the false amnion imitated the positions of the developing embryo (§38). At the end of the third day, the embryo consisted of three parts. In the head region, folded tissue formed a pyramidal head sheath. In the nether region, folded tissue enclosed the tail and feet. Between these two regions, the underbody lay open on the yolk, connected with the yolk membrane by the yolk stalk (§§40–42). Tissue of the false amnion enveloped almost all of the embryo. Although wholly a product of the lower membrane of the *area vasculosa*, the false amnion was equipped with vessels separate from those of the *area* (§44).

In the second and third treatises, Wolff described the transformations of the false amnion between the second and the fifth days of incubation, first as observed from outside then, after some sectioning, from inside. As an armature for his observations, Wolff took the tripartite division of the developing embryo: the head, the midsection, and the tail section.

Seen from outside, the false amnion on the second day of incubation took the form of a pocket consisting of four folds. The first to appear was a head sheath, wrapped tightly around the head and the upper surface of the embryo. During the second day, the head sheath was an oval cavity marking the origin of the *fovea cardiaca* (pt. 2, §§50–57). The *fovea* was to become, as later formation showed, the origin of the stomach. Parts of the false amnion forming the *fovea* subsequently passed over into the stomach. By the end of the third day of incubation, the head sheath had become a complete sack.

At the end of the second day or the start of the third, lateral parts of the midsection lay on either side of the backbone. Their outer limits were folds of the false amnion. Their separate fore edges gradually moved

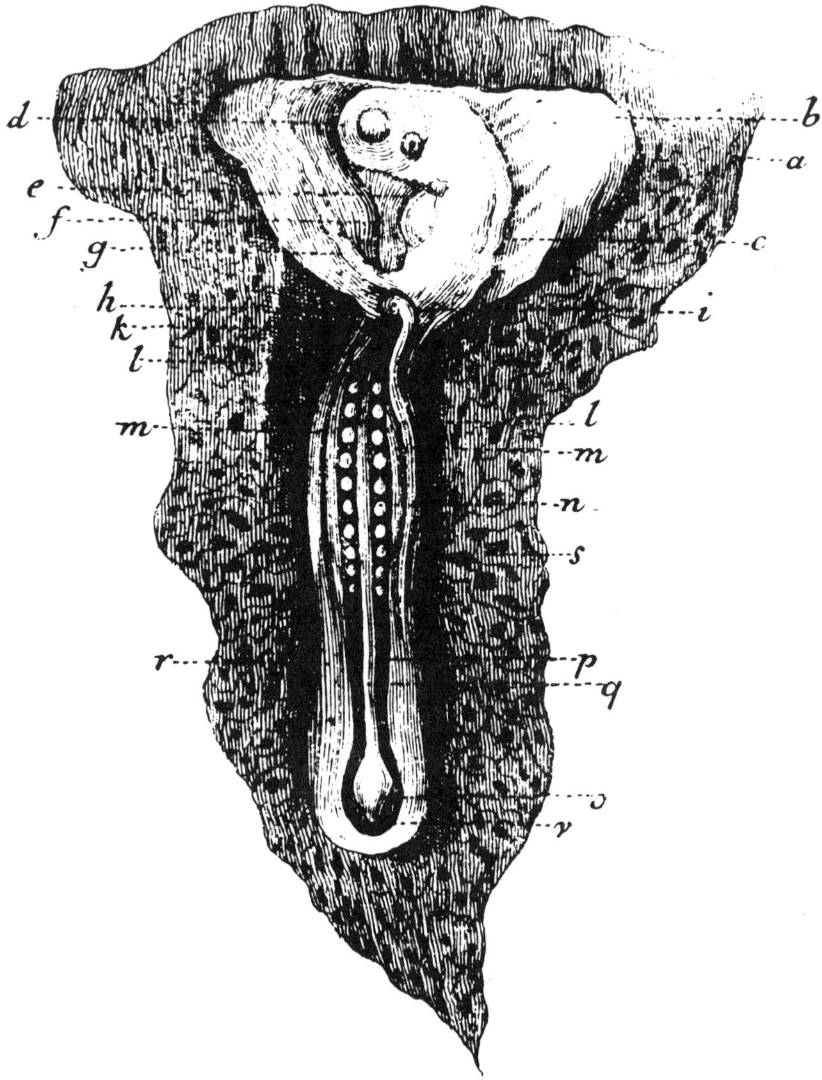

The chick embryo at fifty-four hours, as seen by C. F. Wolff: b, *head sheath;* d, *fore part of the head;* e, *the heart, very contracted;* i, *the swollen edge of the* fovea cardiaca; k, *the* fovea *not yet closed, but narrowing; the swelling midsection;* m, *intestinal swelling;* n, *origin of the spinal column;* u, *rudiment of the tail sheath. Plate 1, figure 5,* Caspar Friedrich Wolff über die Bildung des Darmkanals, *1812, trans. J. F. Meckel (Halle: Rengerschen).*

together to form a suture, which later became the villus of the intestine. The suture first appeared near the *fovea cardiaca*, and gradually extended downward (§§58–64). A fourth fold in the false amnion, whose latter end was the tail sheath, simultaneously formed a lower cavity. This cavity was the origin of the rectum, just as the *fovea* was the origin of the stomach, and the suture was the origin of the midintestine (§§65–73).

This, then, was the formation of the intestine in its earliest period: stomach and rectum were alike; the midintestine was only a little different. A nutritive canal and the upper part of the stomach formed a common cavity. Only a small part of the rest was present, but this part became the basis for all the rest. The front was missing, the stomach was open. The back passed over into the midintestine; again, the back was all that existed. This in turn passed into the back of the rectum, whose front was missing. Only the lowest part of the rectum, like the nutritive canal in the head sheath, was closed (§74).

The literature on generation was full of organic bodies lacking complete parts. Tadpoles, for example, lacked feet, as in the plates by Rösel von Rosenhof. Réaumur and Trembley had discussed regeneration in crabs and polyps. Wolff's own work had demonstrated the existence of a backbone prior to the sprouting of feet and wings. But all of these were whole parts, not indeterminate pieces. Nowhere in the literature was there a description of a part that originated as an open, extended membrane that then rolled up to form a closed canal. This, however, was precisely what took place in the formation of the intestinal canal (§75).[44]

At the end of the second day of incubation, membranes on either side of the spinal column thickened and deepened. The thickest parts began to form a girdle. After the wings sprouted, the girdle was seen to comprise two separate lines, or bulges. The outer line formed the limit of the midsection. The inner line (actually a separate membrane sprouting from the spinal column) developed into the membranes of the intestine. At this point, the midsection of the embryo seemed to consist of two half tubes, one within the other. Then the inner membrane, which was quite flat, struck out over the outer membrane (pt. 2, §§91–104).

At the *fovea cardiaca*, the fore edges of the separate tissue began to pull together, a movement that passed downward to the side pockets. In closing, the borders on the two sides did not pull equally. The right side pulled toward the left, passed under it across the swollen midsection, and reached the inner membrane. Obviously, the closing of the entire opening was subject to a single set of laws (§§104–8).

Closure was only part of a complex series of changes: the embryo contracted and turned on its side; the true amnion began to form; the upper and lower membranes of the *area vasculosa* pulled apart. By the end of the fifth day, the observer found himself confronting an entirely new situation. By this time, the false amnion had revealed itself as a temporary receptacle. The pocket formed by it on the second day of incubation as an inner covering for the embryo developed into an outer covering and was replaced by the true amnion. To this point, Wolff had established that the intestinal canal originated as a simple straight canal, open at the front. Seen from outside, the canal consisted of the *fovea cardiaca*, a midchannel, and the lower cavity. To observe what lay behind this surface, Wolff had to cut away parts of the enveloping pocket. He devoted his third treatise to the results.

Part three was not just a confirmation of his earlier observations, however; Wolff went on to disentangle the origins and formation of the true amnion, the breast cavity, the mesentery, the liver, the kidneys, and so on. The importance of this new material for Wolff's proof of epigenesis is most obvious in his comparison of an embryo at two and four days. Let me use his observation of the heart as an example.

Earlier, Wolff had analyzed the incomplete state of the heart in an embryo of three and a half days. He now began his description much earlier and included all of the factors contributing to the transformation of the heart's location. In an embryo of two days, the breast cavity had not yet formed. Only the back part of the right and left auricles was covered by the true amnion. The fore surface of the auricles, the ventricles, and the aorta lay open on the foremost surface of the breast section of the spinal column (pt. 3, §133). This situation lasted only as long as the heart remained contracted. When expanded by blood, the heart doubled in size and thrust forward "like an arrow." At this point, the heart was covered only by the false amnion. It did not lie inside the embryo. It was not even covered by the embryo's nearest receptacle, the true amnion. This, by the way, refuted Haller's claim, that the heart was at all times protected by the breast cavity.

By the fourth day, plates sprouting from the spinal column began, through a complicated series of developments, to form the breast cavity. This development was related in part to the simultaneous contraction of the embryo—but only in part. "I am convinced," Wolff wrote, "that these changes in the embryo are not the sole or even the main cause: without them the breast cavity would be formed anyway. For to such mechanical causes I ascribe a very subordinate influence on the formation of organic bodies, whose true cause, in my opinion, is to be sought in the forces dwelling in matter" (§135).

The third part of *De formatione intestinorum* included far more than a description of the gradual development of the intestinal canal. Wolff's analysis broadened to include most of the organs formed from materials secreted by the spinal column. Finally, he broke off and turned to his summary argument.

When we observe a foetus of twenty-four hours, he wrote, we find the rudiments of parts that must be transformed into completely developed organs. In the place of the midintestine, for example, located between the duodenum and the rectum, we find two simple lamina that no one would confuse with a complete intestine. One can only conclude that these parts are not yet complete and, therefore, do not always exist but will be formed at a determinate time after conception (§155). Georges Canguilhem remarked that Wolff was unusually sensitive to the novelty of topological structures in a conformation and always resisted descriptions in terms of such ordinary geometrical transformations as *enlargement* or *deformation* of elastic solids.[45]

Wolff's primary proof of epigenesis was different: various parts of the body could not have existed at certain times because other parts occupied their places. The true amnion, a frontal continuation of the foetal trunk, took the place of the breast cavity and the pelvis; in place of the lamina of the midintestine were the leading edges of these lamina; and this was true of the rectum as well. These parts did not exist—not because the observer did not see them but because they did not and could not exist. "On this basis we can conclude that parts of the body do not always exist but are produced successively, however that production takes place" (pt. 3, §155). This was a major concession in terms of the methodological canons that can be traced through most of Wolff's work. "One truly explains generation," he had written at the beginning of his career, "who deduces the parts and the mode of their composition from known principles and laws."

But Wolff could not provide the principles and laws operative in the formation of the intestine. "I do not say," he wrote, "that the formation takes place through the concourse of particles, through the mode of fermentation, through mechanical causes and effects, but only that it does take place" (ibid.). There were complexities that lay beyond his deductive procedure. Rational explanation had, in some cases, to give way to mere description.

This escape from the willed rigor of his method liberated Wolff's observational genius. *De formatione intestinorum* is the most brilliant and imaginative of his treatises. Take, for example, his reflections on the modes of formation in the animal body. At the beginning of the treatise, he mentioned the complexity of the animal body in which various principles triggered various activities to produce the results realized in the different parts. "Among them," he noted, "we find no relation such that one is the basis for another. It is as if diverse natural products, yes, diverse organic bodies, were joined to one another and tied by a band in one body" (pt. 1, §10).

He returned to the idea in the middle of the second part. Although there was no single generational principle at work in the animal body, a single analogy seemed to be active in the formation of its systems. At various times, various systems seemed to be produced according to one and the same norm, and out of these a whole animal was composed (pt. 2, §78). It was as if these systems were alike yet different in principle: all were formed according to a single norm, each as a complete, self-enclosed system. The same act seemed to be repeated several times in the generation of the animal: the various systems were alike in outward appearance but differed totally in nature. Obviously, the causal sequences eliciting comparable acts of formation must differ greatly from one another. "Many diverse principles produce many diverse systems, which, whatever the course of events, are formed according to a single norm" (§79).

Animals, Wolff concluded, were not composed of individual parts but of systems of parts—the medullar, the muscular, the vascular, and the digestive systems, for example. In an adult, the likeness of these systems was not obvious; each was complete and self-contained, an independent formation. One system could not be taken for part of another, because they had undergone changes of figure, magnitude, and disposition that made it impossible to see in the results the likenesses observed in the embryo (§80). Wolff considered this elaborate analogy highly remarkable, "deserving in the highest degree of the attention of physiologists. We must concede that this [analogy] has profound significance and stands in the closest relation with the generation and nature of animals" (§78).

Not all of Wolff's innovations were on so grand a scale. *De formatione intestinorum* contains many insights developed by J. F. Meckel and his generation in the context of comparative anatomy. In describing changes brought about by the closure of the embryo, for example, Wolff remarked that observation was made difficult by the fact that eggs did not mature at the same rate and that not even all parts in one egg matured at the same rate. Meckel applied the remark to anomalies of formation and, in a long footnote in his translation of Wolff's treatise, wrote a short paper on so-called monstrosities.[46]

Another source of interest throughout *De formatione intestinorum* is the polemic against mechanical medicine, the system of *evolutio*, and Albrecht von Haller. Wolff accepted some of Haller's terms for individual parts—the venous figure, for example, and the auricular canal. Otherwise, Wolff took pains to point out the inadequacies of mechanical explanations, the relative maturity of the embryos observed by Haller, the deficiencies in Haller's conclusions, and the wholesale use of invisibility as a source of preformed parts.

Wolff's provocative behavior did not pass unnoticed. Haller reviewed *De formatione intestinorum* in the *Göttingsche Anzeigen*.[47] He contented himself with a summary and added that what seemed simple and flat under the magnifying glass might have been a closed but transparent intestine. Beyond this there was no answer from Haller, nor from anyone else. Wolff's later treatises on embryology and teratology were as good as buried in the memoires of the St. Petersburg academy.

In 1781 the brash young ordinarius at Göttingen, Johann Friedrich Blumenbach, published *Über den Bildungstrieb und das Zeugungsgeschäfte*, in which he cast himself as the champion of epigenesis. Blumenbach did not rest his case, he said, on observation and experience, "the sour fruits of tedious investigation" (p. 7); he offered instead an amusing if glib summary of the absurdities of the doctrine of *evolutio*. Ovism, spermism, and panspermism were weighed against telling, established truths and were rejected out of hand. On one critical point, Blumenbach claimed to have taken a step beyond his predecessors, the men who had moved the science of generation from under the aegis of metaphysics but who had repeatedly found themselves unable to account for the formation mechanism. Blumenbach solved this problem with a freshwater polyp that directed its own regeneration by means of an action proper to its own body.

The point on which Blumenbach's mechanism turned, and which distinguished it from the *plastic force* of the ancients and the *essential force* of Wolff, was *Bildungstrieb*, or *formative force*. He defined it thus: "in all living creatures, there is a particular, innate, lifelong, active, effective force (*Trieb*) that confers a determinate form, afterwards preserves it, and when this is deranged, where possible, restores it" (ibid., p. 12). Blumenbach used the term *force* to introduce the idea of causality acting by design in vital processes. The term was not explanatory, of course; it designated an observed effect. Adoption of the principle was justified by its simplicity and its harmony with observed vital processes.

What, exactly, were the processes Blumenbach had in mind? There was, first of all, the act of fertilization. The male and female prolific fluids, determined for one another at the creation, furnished the material for a new formation. After the fluids had combined, there was a pause, during which the fluids mixed and formed internal relations. This pause constituted a stumbling block for the advocates of *evolutio*. In spite of statements by Hippocrates and Croune, no organization in these fluids had ever been detected (ibid., p. 53). Blumenbach offered his own observations of *Conserva fontinalis L.* Neither he nor C. W. Büttner and G. C. Lichtenberg had detected any sign of a preformed germ under the loupe (ibid., p. 41). After the reproductive fluids had matured, however, the formative force was stimulated (passive voice was obviously created for situations like this) and initiated the organization of previously unformed material. The new formation took place with astonishing speed and always according to kind.

Because the formative force initiated a new formation, Blumenbach found it possible to dispose of many of the problems of deviant formation that evolutionists found so troubling. In very little space, he assembled the various kinds of anomalous formations—monsters, hybrids, races, permanent variations—and showed that all could be explained by the formative force. Like any other physical force, the formative force was subject to physical circumstances. The appearance of a deviation supposed a deviation in the formative force, which in turn presupposed a deviation in normal physical causes.

Not only did the formative force "vivify and shape formless spermatic matter," it preserved "this structure during life, and regenerated any part accidentally mutilated" (ibid., p. 12). This was the source of regeneration in polyps, anemones, and starfish, a capacity having nothing to do with preformed germs. Blumenbach offered two experiments with polyps in which no new matter was produced but in which mutilated matter was restored—a regeneration analogous to the ordinary replacement of matter and the healing of wounds. No doubt the process was simpler in lower animals. The higher animals were composed of many kinds of matter, and the process of regeneration took longer and was increasingly imperfect.

Caspar Friedrich Wolff's reaction to *Über den Bildungstrieb* is unknown. Within the year, though, Wolff had arranged a prize competition in St. Petersburg. His prize question defined nourishment as a process whereby nutrients were brought to all parts of the body for repair and growth and new parts of the embryo were formed. Because this occurred in both plants and animals during development, it could not be ascribed solely to the action of the heart and must therefore be produced by a force. Wolff's question concluded:

> What is the nature of this force? In the first place, is this the same as the universal force of attraction in bodies? . . . In what way is it . . . different from this and proper only to living animal substance and to the vegetable substance of plants? If this latter is true, it is asked further, what are its particular effects, and by what properties is it distinguished from the universal attractive force and its singular and special nature made evident?[48]

Blumenbach submitted an entry of nine pages, half of which were empty posturing.[49] When at last he came to the point, which was after all the central issue in the doctrine of epigenesis, he postulated two forces: the capacity of vessels to drive along and pour out fluids in determinate directions and the capacity of the parts being nourished to attract and assimilate homogeneous parts in these fluids. The existence of these two forces was to be concluded on the basis of observed effects and analogy (p. 12). Whether these forces were separate kinds or merely modifications of one and the same force, Blumenbach left undecided. He inclined to see them as separate (p. 14). In his conclusion, he covered all points of the compass: perhaps his forces could be seen as supports for the well-known essential force, a force to which he now ascribed on the basis of "hundreds" of observations; but his forces were also vital forces, and of course at bottom they were also part of the universal force of brute matter (p. 15).

The St. Petersburg academy also received an answer from C. F. Born, who argued that the basis for movement in nutritive fluids lay in the fluids themselves. The two answers, Wolff noted, contradicted one another, and yet both contained correct observations. Wolff divided the prize between Blumenbach and Born, explaining his decision thus: "The two sharp-sighted authors have, each in his own way, made correct observations. The apparent contradiction is easily overcome and the observations on both sides are easily united. Of course the academy would not crown both works if there were true contradictions in their results."[50]

Wolff took the opportunity provided by the prize contest to publish his solution to the problem of formation as an answer to Blumenbach and Born. The exact nature of the formation mechanism had plagued epigenetic theory since its reform forty years earlier, and for Wolff the outline of the problem had remained the same. Formation was essentially a matter of the apposition of material particles moved by the forces dwelling in matter. Wolff was prepared to admit that there were complex modes of formation beyond the explanatory potential of this mechanism, modes that could be observed and described but that were not related to any known explanatory principle. And he was prepared to concede to purely physicomechanical forces a subordinate role in formation. But he continued to claim that mechanical forces could not account for the movement of fluids in the animal body before the formation of the heart. Blumenbach's use of a formative force, on the other hand, was nothing more than an admission that the organism could not be derived from known physical causes (ibid., p. 66n).

Wolff proposed a model for formation that was mechanical in its own right and yet peculiar to living creatures. He had not reformulated the problem of formation and could not, by his own admission, account for his own most brilliant insights, but he did propose a comprehensive analysis of the difficulties and implications of an intelligible formation mechanism that surpassed all earlier statements in thoroughness and lucidity.

Wolff's thought circled around the modified version of the force of attraction found in the chemistry of affinities. The essential force was attractive and expansive; its effects, however, could be various. They included, as in the chemistry of affinities, both attraction and repulsion. In the course of vital processes, these opposed effects were further modified by physical contexts and other causal factors. Nourishment offered a simple illustration of the operations of this force.

Nourishment was essentially a matter of the circulation of fluids by means of vessels to the fast parts of the body. The fast parts of the body attracted like materials, for maintenance and growth, and repelled unlike materials. The basis for both attraction and repulsion was the force present in both the fast parts and the fluids. They attracted one another, Wolff wrote, as do magnets and iron. The same force was also present in the walls of the vessels through which the fluids flowed; the movement of fluids was operated by the mutual repulsion of fluids and vessels. What Blumenbach had taken for two forces, said Wolff, was in fact one and the same force, producing different effects all according to different circumstances (p. 67n).

During organic formation, parts solidified sequentially. As they did so, they expelled dissimilar materials, which went on to form other parts. All parts continued to grow by attracting new materials. In an incubated egg, for example, the process was initiated when the yolk grew warm and dissolved into fluids. These fluids, developed for the formation and nourishment of the embryo, repelled and were repelled by the yolk. At first the fluids seemed to lack direction, something they acquired only through attraction. They then flowed to the site of the embryo, where their movement laid down the rudiments of the vessels in the *area vasculosa*. As vessels took shape, ramified, and cohered with the embryo, the fluids expelled from the yolk attracted and were attracted by the openings of the vessels and began to flow in determinate directions by purely mechanical means. Only at this late date did the essential force take on the mechanical causes and effects that would henceforth modify its actions so greatly.

Wolff was quick to add that the basis for all life and vital processes in plants and animals could only be the essential force. The initial movement of fluids in the incubated egg was based on attraction and repulsion, long before the existence of any circulatory apparatus or the formation of the heart. Mechanical causes, such as peristalsis or the pumping of the heart, appeared only later. They influenced vegetative processes but did not cause them.

Suppose, Wolff argued, that one could construct an artificial *Tragopogon* from matter subject to the ordinary force of attraction. And suppose that this model resembled the *Tragopogon pratense L.* down to the least detail, not just outwardly, but also inwardly. Would this machine grow, flower, fruit, and reproduce? "I think this model would remain the same," Wolff wrote, "and I think that not even the most zealous partisan of mechanical medicine would attribute these processes to the model." The *Tragopogon pratense L.* was alive because of the essential force, and this force must have been peculiar to plant and animal substance, since only this substance was nourished, vegetated, and reproduced its kind." And since, moreover, the whole life of plants [and animals], their growth, nutrition, vegetation, and reproduction, depend upon this force, we may call it a peculiar and essential force" (ibid., p. 39).

Although it was peculiar to animal and vegetable substance and the basis for all vital phenomena, the essential force was not a vital force, according to Wolff. The essential force seemed to be "that for which Albinus sought, and whose existence Stahl recognized but ascribed incorrectly, in my opinion, to the soul" (p. 42). The basis of life was not the soul; it was the particular and determinate force of attraction and repulsion characteristic of animal and vegetable substance.

The essential force was not to be confused with the formative force, either. Blumenbach's *Bildungstrieb* revived all of the absurd and embarrassing claims made on behalf of forces since the time of Maupertuis. A simple natural force produced a simple unvarying effect. For this reason, the formation of complex structures could not depend immediately upon one single force. "There would be no reason why this force would work now in one way, now in another, nor why it would produce at one time one part and at another time another" (p. 66n). Blumenbach's formative force could not be demonstrated scientifically. If such a force were to form a plant and produce a root on one side, "it could not produce a stem and leaves on the other side. For it is still the same force, and it must have the same effect. If in these two different results there were the least difference, this would be entirely without a reason" (p. 67n).

Wolff's essential force was, by way of contrast, a simple natural force, whose effect was always the same. Like substances attracted like substances and repelled unlike substances. The action was sufficient to distinguish the essential force from other kinds of surface attraction: it was peculiar to plant and animal substance and, thus, characteristic. As the basis for nutrition, it was also essential—hence, the characteristic and essential force of plant and animal substance.

The force attracted or repelled according to circumstances, and these circumstances modified not the force but its effects. The essential force,

Wolff wrote, "does not produce the various parts of the body through itself and according to its nature [as in the *Bildungstrieb*] but with the aid of innumerable other concurring causes; and what the essential force does on its own is always a simple effect, as attraction, or repulsion, worlds apart from the formation of organic bodies" (p. 64). What was attracted or repelled did not depend upon the force alone but upon the circumstances in which it acted. Together, the circumstances and the force were sufficient to produce all of the processes of organic vegetation.

It proved impossible to specify exactly the nature of the likenesses and differences that triggered attraction and repulsion. In one context, Wolff's model seemed to be chemical affinities; in another, the model was hylozoism; and in another, it was Stahl's animism (p. 67n; see also pp. 35–36, 43–44, 53–54, 59). Wolff found his most persistent analogy in the universal force of attraction. Perhaps, he remarked at one point, all natural occurrence can be reduced to attraction and repulsion, and perhaps these in turn depend upon a single original force.

> In all nature but one force seems to exist, namely the force of attraction and repulsion, and this is differently determined in different substances. Thus there are not two forces but only one, and all the differences that exist in the effects of nature, if they are not merely mechanical modifications of these effects, depend upon determinations of this single force. (pp. 69–70)

This uncharacteristic bit of *Naturphilosophie* was only one of the ways in which Wolff the rationalist looked forward to the world of Schelling and Treviranus.

Wolff' s attempt to provoke a controversy with Blumenbach misfired. By 1789, when Wolff's *Streitschrift* was published in St. Petersburg, Blumenbach's work had initiated a new period in the science of generation in the German-speaking world. Blumenbach's combination of physicomechanical and teleological principles in the formative force acquired influential supporters. "Your work," Immanuel Kant wrote Blumenbach, "has taught me a great deal, particularly in the union of two principles, namely the physicomechanical and the teleological kind of explanation of organized nature, which one had previously believed incompatible."[51]

With this powerful endorsement, a vitalist school of physiology grew up at Göttingen, with Blumenbach at its center. "Vital forces," he claimed, "are merely terms given to effects known a posteriori, such as the *Nisus formativus*."[52] He had adopted the term, he said, to distinguish it from other orders of vital forces and by no means to explain the cause of generation, which he considered as involved in Cimmerian darkness as the cause of gravitation or attraction. Blumenbach went on to warn that any attempt to define the *Nisus* more closely would only obscure the problem.

In spite of his warning, many did attempt to define the formative force in the 1790s.[53] Life was the capacity of matter to act out of an inner principle of thought, the activity of matter according to the laws of organization, the capacity for motion at the service of the moved. "Living bodies," wrote Alexander von Humboldt, "are those that in spite of incessant attempts to alter their form are hindered by a certain inner force from abandoning their first characteristic form."[54]

Since the formative force was determined solely through its effects, an observer tended to concentrate upon the description of those effects. Freed from the burden of offering a sufficient reason for every embryonic transformation, physiologists turned instead to the discovery of analogies between individual formation and other natural processes. An early and important analogy was the concept of recapitulation. As Blumenbach's student C. F. Kielmeyer explained it,

> Since the distribution of forces in the series of organized beings follows the same order as their distribution in the developmental stages of given individuals, it follows that the force by which the production of the latter occurs, namely the reproductive force, corresponds with its laws to the force by which the series of organized beings of the earth were called into existence.[55]

We have no reason to believe that Wolff would have felt uncomfortable with Kielmeyer's vision. In his sober, disciplined way he had laid a basis for just this insight thirty years earlier. But was Kielmeyer writing science? Wolff would have objected vigorously to a discourse so thoroughly lacking in verifiable principles and morphological detail. After *Nutritionskraft*, however, Wolff published nothing more. He continued the quiet round of his research until his death in St. Petersburg in 1794. At the time, he was remembered by German contemporaries, if at all, as the opponent of Albrecht von Haller and Charles Bonnet in the old quarrel over *evolutio*.

J. F. Meckel published a translation of *De formatione intestinorum* in 1812. The treatise had lain forgotten for fifty years. Meckel implied that Wolff's reputation had suffered the same fate. The world into which Meckel reintroduced Wolff seemed to have abandoned close reasoning and patient observation—so much at least we can surmise from Meckel's praise of Wolff's qualities as a scientist.

> This close observation, this unwearying step-by-step pursuit of an organ from its first formation to its most complete development, without being

> led by preconceived opinion, without saying more than one sees, without establishing laws out of what are only probable conjectures: this alone can lead to a true developmental history of the embryo, and cannot be sufficiently commended or encouraged.[56]

Meckel used Wolff's treatise as a corrective to the assertions of Lorenz Oken, Emmert, and Hochstetter. As a comparative anatomist, Meckel also pointed out Wolff's contributions to the laws of form—the idea of a unity of plan, for example, the use of anomalies to illuminate normal development, and the parallel between embryonic development and the animal series. All of this fell on fertile ground. After 1812 the citations of Wolff were rather numerous.

Revival of the old and discredited system of two fluids promised, as the system of *evolutio* did not, to establish and maintain a firm connection with the efficient causes and physical processes that made organic formation intelligible. Once restored to a proper physical context, the process of generation would disclose its own characteristic regularities and suggest analogies between these and other physical processes.

In effect, the program of Maupertuis and Buffon amounted to a reevaluation of the factors of heredity and environment in organic generation. At the level of individual formation, Maupertuis equated intelligibility with the time that succeeds itself and interested himself in the transmission of abnormalities like polydactyly or mutilations that lent themselves to observation and experiment. Buffon interpreted intraspecific diversity as a degenerative response to physical conditions; at the species level, he substituted reproductive criteria for abstract affinities. Quite consistently, he went on to argue that groups of species were united by common historical origin and to conceive the possibility of a history of life. All of this followed from the programmatic stress on the succession of living forms in relational time and space.

The ambitions of Maupertuis and Buffon count for more, however, than their actual achievements. They promoted observation and experience, but neither man availed himself overmuch of these resources. Let me mention one last time their failure to find a female prolific fluid and their use of attraction as the agent of formation. These were not incidental lapses.

These embarrassments were only partially obscured by the rhetorical stances of the two principals. Maupertuis offered a gentlemanly ramble through the curiosities of generational theory. He presented his "vague thoughts" and "pleasantries" as subjects for amused speculation. Buffon, in a more ponderous way, insisted upon a methodological distinction between preliminary statements and finished doctrine. Some critics may have been baffled or disarmed. In the end it was Albrecht von Haller who brushed aside these sophistries and focused attention upon the dangerous novelties in the basic concepts of matter and formation. And it was Haller who insisted that the science of generation, a difficult and intricate discipline, was not a subject for frivolous speculation and profane conjecture. The serious implications of the development of living beings were not well served by worldly wit and trifling inferences.

Haller set up his observations of incubated eggs on a schedule that went around the clock. He pursued the complications of vessels and membranes with unrelenting tenacity and pushed as far as his limited visual means allowed. Away from his microscope, Haller ransacked the literature on generation, pursuing possible interpretations and arguments for his campaign. The record of his observations, *Sur la formation du coeur dans le poulet*, is one of the great monuments of observational science in the eighteenth century. Haller succeeded where Maupertuis and Buffon had failed, as a scrupulous observer with an unparalleled grasp of anatomy and physiology.

But his observations, good as they were, were subject to correction. Anyone who reads Caspar Friedrich Wolff attentively will find more than a hint that Haller began his observations of the embryo rather late, that he invoked the initial invisibility of parts too often, and that he tended to see what his convictions allowed him to see. Wolff put his finger on a central problem: Haller could be blinded by his own beliefs. Haller's embrace of the theory of *evolutio* was a conversion. Thereafter, he defended the doctrine with all of the resources of his great authority. His opinions did not pass unchallenged, however, and the effect of his arguments on the science of generation was less severe than in other areas of physiology.

In another way, too, Haller's subversion of the epigenetic program failed. For purely tactical reasons, he had based his early arguments for *evolutio* on the regularities and continuities of physical development. His assumptions and conclusions could be questioned, but his detailed representation of generation as a relatively self-contained developmental cycle helped liberate generation from the territorial claims of metaphysics and theology.

It is almost impossible to do justice to Wolff's contributions to the science of generation without exaggerating his influence. His appointment in St. Petersburg took him outside the German-speaking world, and his efforts to reestablish contact with that world failed. It would be possible to write a history of the theory of generation in Europe during the later eighteenth century without mentioning Wolff, but it would be a rather superficial history. In spite of his isolation and his inability to persuade his contemporaries and immediate successors of the importance of his ideas, Wolff's work clarified the epigenetic program, and what he clarified proved central to the scientific culture of the later eighteenth century.

Wolff's early work demonstrated how parts necessarily formed in serial order through the production and organization of structure. Many of the concepts Wolff defined so scrupulously—growth, nutrition, vegetation, organization—were borrowed from tradition. The process of generation implied gradual formation, but this was designated as *compositio partium*, *structura*, and so on and was not closely specified. Wolff also mentioned an essential force that provided the sufficient reason for all vegetation in plants and animals. This force may have been, as Georges Canguilhem has remarked, the least original of Wolff's many *êtres du raison*,[57] but this force also offers the historian the best access to Wolff's assumptions.

The theories examined in this study shared the same mechanist conception of formation. The simple elements, ingredients, and principles of the system of two fluids lacked organic character and required an agent to operate formation. The conception foundered repeatedly. Simple natural attraction, however varied and modified, was not up to the complexities of living form and could not account for the unerring achievement of structure. Wolff's essential force in its final version—a complicated and subtle modification of affinity—shared the fate of earlier attractive, expansive, penetrating, and repulsive forces.

Wolff pushed beyond these assumptions in *De formatione intestinorum*, offering a brilliant study—purely descriptive, it is true—of the formation of the intestinal canal. At the time, his feat of imagination was ignored. The science of generation went off in another direction. Epigenesis did, of course, became the dominant theory in the science of generation, but it was not the epigenesis of Caspar Friedrich Wolff. Mechanist epigenesis was discarded along with the system of *evolutio*. Neither theory was disproved; they were simply rejected as the sour fruits of tedious investigation.

Blumenbach's proposal of a formative force commended itself to many by its simplicity and its correspondence with nature. As a directive force, added to effects known a posteriori, it provided a way of unifying the natural processes of growth and reproduction. The formative force was not observed, of course; strictly speaking, it was not observable. But its effects were evident. With this conception in hand, physiologists could take the fact of organization for granted; they did not have to explain it.

In retrospect, the epigenetic program launched by Maupertuis and Buffon and carried out in a very different way by Wolff seems to have come a cropper. Wolff's grand science of generation, comprehending anatomy and physiology, had no recruits. The materialist explanation of formation, along with its mechanist models, was discarded. We seem to be confronting complete failure.

But if we return to the point where we began, we find at least one issue on which the proponents of epigenesis enjoyed success. They represented organic formation as a coherent and intelligible physical process, unfolding without divine intervention. Vitalist physiology took over this element in the older epigenetic program without a murmur. And when after a time young vitalists like Gottfried Reinhold Treviranus discussed the spiritual aims of the universe and the hyperphysical character of life, they took it for granted that these were realized in the succession of time and through the forces dwelling in matter.

VI

Kant and the Critique of Teleological Judgment

In 1790, at the age of sixty-six, Immanuel Kant published the *Kritik der Urtheilskraft*, an analysis of the principles of judgment which he situated between the realms of nature and freedom.[1] Obviously a product of the special demands of the Kantian system, the third critique also gave expression to some of the most profound intellectual and spiritual tendencies of the late eighteenth century. The book's effects were felt in many fields of cultural activity—the arts, criticism, philosophy, theology, and of course, the life sciences.

This discussion limits itself to the second part of the third critique, the critique of teleological judgment, and concerns itself only with arguments that deal in one way or another with the attribution of purposiveness to living form. The study excludes as far as possible any attention to the third critique's relation to Kant's philosophy as a whole, to its internal architectonics, and to its analysis of the theological abuses of teleology.

Kant's name has cropped up repeatedly throughout these studies. He was a shrewd student of the life sciences and had followed their development since the 1750s. As a professor he pursued and occasionally contributed to the fashionable science of ethnography. Kant was one of the very few who followed—and understood the implications of—Buffon's distinction between abstract and physical truth. This much is clear from the lectures on physical geography initiated in 1757 and from the three papers on the human race discussed earlier. Kant distinguished even more clearly than Buffon between the atemporal taxonomy of contemporary natural history and the genetic history of nature. He made no secret of his preference for the latter, and for this reason his opinions were considered unworkable or superfluous by working naturalists.

In the critique of teleological judgment, Kant not only repeated his old arguments for an alternative physical history of nature, but he analyzed his critics' basic assumptions about living form and natural order. In line

Immanuel Kant (Raab's engraving of the portrait of Döbler)

with the fundamental tendency of critical philosophy, Kant was not so much interested in natural order as in the concepts with which naturalists and physiologists represented that order. The results of his analysis were devastating: he argued that, for all the emphasis upon purely physical explanations, teleological judgments remained the conceptual center of the sciences of living form.

Naturalists and physiologists continued to regard the living kinds as superior artifacts, designed to take form and to function independently, while contributing to the economy of nature as a whole. They considered the economy of nature to be regulated by a priori principles, which maintained nature's balance and harmony. They represented the formal structure of nature as a plan disposed according to the law of continuity. In each of these cases, Kant detected an analogous conception of the relation between a whole and its parts that drew upon the broad definition of purposiveness in contemporary usage. A whole was said to be purposive when its parts were not merely juxtaposed but were mutually dependent. The whole formed a system whose parts were differentiated functionally; their functions harmonized in a unified, concerted action with a single, overall significance. In the economy of nature, for example, every living thing was determined to some certain end, and nothing was unnecessary.

Kant did not propose to eliminate these kinds of teleological judgment outright, whose value as heuristic devices he understood very well. But he did insist upon some recognition of the limits to their validity. Judgments of formal purposiveness found a certain justification in reference to the immanent structure of living nature but not as signposts pointing to an external, transcendent ground on which nature depended.

In the original introduction to the third critique, which he later suppressed because of its length, Kant analyzed the tie between scientific assumptions and the system of nature. His analysis is, or ought to be, at the center of any discussion of eighteenth-century systematics.

The rules and forms of classification with which naturalists mastered the diversity of nature were determined in part by a concern for a purely descriptive study of nature. The autonomy of the sciences of natural history manifested itself in the attempt to grasp living nature as a whole, self-sustained and self-explanatory. If, in some respects, older metaphysical notions still betrayed the ambition to explain nature, naturalists nevertheless agreed that the order of things was best approached through observation of the regular disposition of forms in nature. Their assumptions were so simple, and agreed so thoroughly with the fundamental conceptions of the ordinary view of the world, that they seemed to offer no foothold for criticism.

Naturalists assumed nothing but the existence of natural forms in all of their diversity and the power of the mind to select from this diversity characters common to some of the individuals. When the naturalist collected forms into classes on the basis of common marks and repeated the same process of abstraction at more general levels, he created an order and division of being based on the likenesses and differences observed in particular forms.

But to Kant there was a discrepancy between this descriptive activity and the validity naturalists derived from it. The grouping and division of natural forms on the basis of affinities did not simply lead to workable categories but, if rightly carried out, was supposed to reflect nature's own plan. The naturalist isolated a determining pattern, an ideal type, active in a manifold of individuals and used superior groups to set forth the basis for each special form. The conceptual hierarchy reflected the process by which substance unfolded in special forms of being. A complete system of scientific definitions was, then, a complete expression of the substantial forces that controlled reality.

The aim of the sciences of natural history was to remain within the world of natural forms and to reveal its constant and universal order. But when one approached nature from this side, there was no guarantee that she was constant or that she even had an order. Classificational methods assumed that the order of nature would emerge from the observation of the regular disposition of forms. But in fact, naturalists' assumptions about natural order shaped their observations. The idea of a system of nature did not follow from experience but constituted a presupposition that naturalists applied to experience and, in a sense, rediscovered there.

This was not a problem to engage a working naturalist. For the critic of knowledge, however, the questionable harmony between a conceptual hierarchy and the order of nature constituted a significant puzzle. The concepts of class, order, genus, and species were logical and a priori, but it was not certain they would be found in nature. In what sense were naturalists justified in describing nature in the form of a logical hierarchy? If the question had an answer, it could lie only in discovering some justification for the procedure.

The form of deduction that Kant had used in the *Kritik der reinen Vernunft* for causality, substance, and other synthetic principles did not apply here. There, Kant had shown that causal and other synthetic principles were necessary conditions for all possible experience. The synthetic principles were universal constituents of our conception of nature. If one thought of nature, one must necessarily think of her as subject to the rules expressed in those principles.

In the *Erste Einleitung*, Kant went on to argue that one could imagine a nature subject to these synthetic principles but that, in other respects, might disclose an irreducible diversity.[2] This nature would obey the rules of connection according to law, but she might not allow us to arrange individuals, by means of affinities, according to class and order, genus and species. A naturalist would find himself unable to compare forms; each natural form would be unique; to be known, it would have to be described for itself alone.

> One might wonder whether Linnaeus could have hoped to design a system of nature if he had to worry that a stone that he found, and that he called granite, might differ in its inner character from any other stone, even if it looked the same, so that all he could ever hope to find would be single things—isolated, as it were, for the understanding—but never a class of them that could be brought under concepts of genera and species. (ibid., p. 216n)

Human understanding could not grasp the possibility, let alone the necessity, of such a whole. Obviously, Linnaeus could not have undertaken his classification of nature if he had not assumed that nature would meet him halfway and disclose a thoroughgoing arrangement in all of her forms. Kant called that assumption a principle of reflective judgment and explained that it was not to be taken dogmatically. It was a procedural rule to guide investigation.

> Surely something is presupposed whenever we compare empirical ideas in order to know empirical laws, and accordingly, the specific forms and, through their comparison with others, the generically similar forms in natural objects: what is presupposed is that nature in respect to her empirical laws has observed a certain economy suitable for our judgment and has adhered to a uniformity we can grasp; and this presupposition must precede all comparison as an a priori principle of judgment. (p. 213)

As an assumption, the principle of reflective judgment was necessarily indeterminate. The naturalist did not know beforehand whether nature had arranged her forms so that all occupied determinate places in a single system. He simply sought likenesses and differences in individual forms that allowed him to class them in genera and species. If one particular group proved unfounded, he replaced it with another; he did not, however, replace the conceptual hierarchy. Thus it was obvious that this framework prescribed no particular content and yet was decisive for the form of descriptive science.

"Judgment orders appearance according to the universal though at the same time indeterminate principle of a purposive arrangement of nature in a system," Kant said, "as if nature were something favorable to our judgment in the conformity of her particular laws . . . to the possibility of experience as a system" (p. 214). In this way, judgment attempted to harmonize the interrelations of living forms with the conditions for the possibility of experience in general. But the formal arrangement of nature assumed by judgment did not prescribe a law to nature; it was simply a heuristic device for judging nature. The aphorisms of descriptive science—that nature always chooses the shortest path, that she does nothing in vain, that she suffers no leap in the manifold of forms, and though rich in varieties is poor in species—were all "transcendental utterances of judgment" (p. 210). Whether or not these aphorisms were reliable, whether or not a complete classification was possible, only nature could decide. But the naturalist would not discover a system in nature if he did not first seek it.

"Without some such presupposition," said Kant, "we cannot hope to find our way in the labyrinth of multiplicity of possible separate laws" (p. 214). The heuristic device rested upon a tacit analogy between natural forms and logical forms. Reflective judgment "constrained" the naturalist to seek the likenesses and differences that would allow him to arrange natural forms as genera and species. The sciences of natural history not only existed on this basis but developed progressively, an indication that the analogy was not entirely arbitrary. Later, in a letter to Beck, Kant emphasized nature's accommodation to the limits of human judgment "in the simplicity and perceptible unity of her laws and in the presentation of the infinite diversity of her species according to a certain law of continuity that allows us to relegate them under a small number of generic concepts."[3] The conception resulted from a special and independent interpretation, which Kant called the technic of nature.

The idea of a technic of nature, of a logical organization of forms, regarded the formal purposiveness of nature as a product of art. "When reflective judgment tries to bring given appearances under empirical concepts of determinate natural things," said Kant, "it deals with them technically . . . in terms of a principle that is universal but also indeterminate: the principle of purposive arrangement of nature into a system—an arrangement, as it were, for the benefit of our judgment."[4]

We assume that nature in her diversity has observed a division into genera and species, he said, and this enables us to arrange natural forms systematically. The unity of natural forms could be regarded as if some transcendent being had created it for the benefit of our understanding: reason seemed to indicate such an idea.[5] But he argued against transforming a mere premise into an objective power in things. Natural order was posited in the course of scientific inquiry and was grounded more thoroughly with every new step. A naturalist might go even further and regard the logical unity of natural forms as if an intelligence had produced it, but nothing obliged him to do so, since judgment in this case was only mapping a course for investigation.

> So if we say that nature makes its universal laws specific in accordance with the principle of purposiveness for our cognitive power . . . then we are neither prescribing a law to nature nor learning one from it by observation

(although observation can confirm the mentioned principle). For it is a principle not of determinative but merely of reflective judgment.[6]

Reflective judgment dealt in the connexity of concepts, not in the order of things.

Kant's solution to the problem contained in the systems of natural history was not the simple methodological adjustment it seemed to be. The principle of reflective judgment did not so much amend the traditional assumption of a harmony between nature and the understanding as it reestablished the assumption in the framework of a grammar of thought. The basic terms of contemporary natural history—*system*, *order*, and *purpose*—continued to play a role in Kant's discussion, but only at a distance. They were regarded and justified as ideas shaped in consciousness. The principle of reflective judgment attributed nothing objective to nature but simply indicated how a naturalist must proceed when reflecting upon the objects of nature with the aim of an entirely coherent experience. The subjective expression of this procedure was the feeling of pleasure that accompanied every discovery of a structural consonance: "We rejoice (actually we are relieved of a need) when, just as if it were a lucky chance favoring our aim, we do find such systematic unity among merely empirical laws, even though we had to assume that there is such a unity although we have no insight into it and cannot prove it" (ibid., pp. 186–88).

Wherever it appeared in the sciences of living form, the idea of finality brought with it a particular kind of relation between a whole and its parts. The principle of reflective judgment, for instance, conceived the laws of nature as a system unfolding according to a rule. The whole was not regarded as the aggregate of its parts but as their source and the basis for their determination. This whole was ideal, an assumption that reflective judgment applied to natural forms and laws; it was not an objective and active power in the forms themselves.

Kant's discussion of the organism in the *Kritik der Urtheilskraft* contained an analogous conception of the whole and its parts. In this case, however, the relation did not seem to be entirely ideal. When anatomists dissected plants and animals, they proceeded as if nothing in the structure of the parts were gratuitous (ibid., p. 376). Without this principle, there would have been nothing to unify their observations. But its application led directly to the recognition of a kind of organization different from that found in other natural objects.

The possibility of the parts in an organism depended upon a relation to the whole that determined everything the organism contained. Moreover, the parts combined into the unity of a whole because they were reciprocally cause and effect; through their own causality, the parts produced one another as regards both form and combination, and in doing so, they produced a whole. "Only under these conditions and upon these terms can such a product be an organized and self-organizing being and, as such, can be called a natural purpose" (p. 374).

Where the idea of purpose does not refer to a relation between our cognitive powers but is perceived objectively, we are dealing with an end of nature as an organized being, that is, an organism. One tree generates another tree, preserving the species. The tree also produces itself as an individual, enlarging and renewing its parts in an orderly way. We call this activity *growth*, but the process involves more than an increase in size according to mechanical laws. The matter assimilated in growth takes on a quality peculiar to the species; its composition is the tree's own product. Moreover, a tree has the ability to generate itself by grafting and to compensate for deficiencies and defects. In short, organisms possess qualities that indicate a principle of formation intrinsic to individuals and the species (pp. 371–72). Kant contrasted the formative principles of a clockwork and a living body to point up the peculiarities of inner purposiveness.

> In a clock, one part is the instrument that makes the others move, but a wheel is not the effective cause that produces another. One part is certainly present for the sake of another, but it does not owe its presence to the agency of that other. . . . Hence one wheel in the watch does not produce the other; still less does one watch produce other watches by utilizing or organizing alien matter. It does not itself replace parts of which it is deprived, nor, if these are absent in the original construction, does it compensate by having the other parts help out, let alone repair itself when it is out of order. Yet all of this we can expect organized nature to do. Hence an organized being is not a mere machine. For a machine has only motive force. But an organized being has within it a formative force, and a formative force that this being imparts to the kind of matter that lacks it (thereby organizing it). Therefore, this force is a formative force that propagates itself—not to be explained by the capacity of movement alone, that is to say, by mechanism. (p. 374)[7]

This statement of the problems presented by the objective study of organisms was preliminary to Kant's methodological analysis. He was engaged in a critique of judgment, not a metaphysic of nature. The question, then, was not whether nature was guided by an alternative causality in some of her products but rather whether an investigator must posit a special kind of natural object, different from the objects studied by abstract mechanics.[8]

Obviously, there was a demonstrable difference between the mechanical processes found in a clockwork and the functions of a living body. Practitioners of mechanical medicine explained the movements of the body on the basis of structure, and their mechanicocausal studies were a legitimate contribution to ordinary theoretical science. But since an organism was not only organized but self-organizing, a mechanist could not completely understand organic form in purely mechanical terms. This was the basis for regarding the form as if it had come about "through the ability to act according to purpose" (p. 370).

At this point, the idea was limited to a particular form, for which nature seemed to contain no basis (p. 422). The observer introduced teleological considerations to compensate for the inadequacies of mechanical methods (p. 383). Blumenbach, Kant wrote, "rightly declared it contrary to reason that crude matter on its own should have structured itself originally in terms of mechanical laws, that life could have sprung from the nature of what is lifeless, and that matter could have molded itself on its own into the form of a self-preserving purposiveness" (p. 424).

Only after Blumenbach had introduced the principle of an original organization was he in a position to "prove" the theory of epigenesis. His principle was a determinate concept, but its application remained reflective. He did not attribute to nature a cause that acted intentionally.[9] The statement, for instance, that the crystalline lens in the eye served the purpose of focusing light rays on the retina guided investigation of the eye and might help devise means to further that purpose, but it said nothing about nature forming the lens with that intention in mind.[10] This use of purpose explained nothing objectively but only according to the range and limits of human understanding; it offered no insight into how organisms are formed but was a maxim of judgment, a guide for investigation. The physiologist who, like Blumenbach, applied teleological principles to organisms drew an analogy with human productivity, and in doing so he ran the risk of confusing nature with intentional creativity.[11]

But organisms remained products of nature, and the physiologist would discover no intentional agent behind them. He might speak of nature as if her purposiveness were intentional, but in doing so he was merely pursuing a causal connection in organic forms; he neither attributed nor denied intentions to nature. "If a judgment asserted either of these alternatives," Kant wrote, "it would no longer be reflective . . . and the concept of purpose would no longer be a mere concept of the power of judgment . . . but would be connected with . . . the concept of a cause that we posit beyond nature and that acts intentionally, a concept that is transcendental."[12] Mechanical principles offered the physiologist the possibility of an objective understanding of the production of things. This was so because we use mechanism to produce things and have insight into mechanical principles.

"When we study nature in terms of its mechanism," said Kant, "we keep to what we can observe or experiment on in such a way that we could produce it as nature does, at least in terms of similar laws; for we have complete insight only into what we can ourselves make and accomplish according to concepts."[13] Organisms may be produced on mechanical principles, and in the future another Newton might explain the production of a blade of grass in terms of laws unordered by any intention (p. 400). But it was clear that, given the present limits of human understanding, we could not become familiar with, much less explain, organisms in terms of mechanical principles alone. It was reasonable for a scientist to proceed in terms of mechanical laws as far as he could, however, "for in these laws lie the true physical bases for explanation, whose coherence constitutes what scientific knowledge we have of nature through reason."[14] Further, one must assume that some original organization (irgendeine ursprüngliche Organisation) made use of mechanism to produce other organisms or to modify its own form.

A complete explanation of organisms on either mechanical or teleological principles remained impossible, of course, although scientists continued to judge organisms in terms of both. Again, these principles, when applied to organisms, were merely reflective. A physiologist judged relations among the parts of an organism in terms of efficient causes, trying to gain as much insight as he could in terms of mechanism, and then judged the same relations teleologically, to discover characteristics that went beyond mechanism.[15]

Judging an organism to be a natural purpose on account of its form was very different from judging the same form to serve a purpose in relation to other forms. But the two kinds of judgment were related. Anatomists assumed that there was nothing gratuitous in the inner structure of plants and animals. Naturalists applied the same principle to external relations among plants and animals. According to Kant, "Everything in the world is good for something or other; nothing in it is gratuitous; and the example that nature offers us in her organic products justifies us, indeed calls upon us, to expect nothing from it and its laws except what is purposive in relation to the whole" (ibid., p. 379). His objection to this way of thinking is implicit in the statement: relative teleology often amounted to a confusion of the idea of finality with that of utility.

Kant's prime example of utilitarian teleology in science was Linnaeus's

Politia naturae, the dissertation in which Linnaeus (on p. 173) explained "to what purpose each has been created and the combination that is among created things in their generation, preservation, and destruction." Linnaeus argued that all nature aimed at one ultimate end and thus connected all natural objects through reciprocal uses. Led by an overt analogy with social organizations Linnaeus concerned himself with the general benefits produced by reciprocal uses. The complacency of his observations was mitigated by the rigor and impersonality of his conception of the natural mechanism. The economy of nature functioned for the good of all but to the complete satisfaction of none. There was a latent opposition between the pleonexia of individuals and the inexorable requirements of order. For the individual, this might mean suffering and death, but the overall result was the balance and harmony of nature.

In considering food chains, Linnaeus advanced what seemed to him a paradox: plants have not been created for insects, nor small birds for birds of prey, but rather the reverse. Herbivores moderated the growth of plants, predators limited the herbivores, and humans hunted predators to establish a balance between the productive and destructive forces in nature. Kant noted that, although Linnaeus had conferred upon humankind the dignity of serving a purpose, he had thereby defined humans as a means. The human species, supposedly the ultimate link in the chain of purposes, was actually, according to Linnaeus (ibid., p. 169), only another servant of nature, dedicated like all the others to maintaining "a seemly balance . . . that nothing becomes too much." Contrary to his stated intention, said Kant, Linnaeus showed that nature had not exempted humankind from her destructive forces "any more than from her productive forces, but had subjected everything to a natural mechanism without a purpose."[16]

Linnaeus's use of teleology was by no means the worst example of widespread contemporary practice. By the mideighteenth century, a utilitarian outlook and an attitude of narrow calculation had trivialized the old idea of a theodicy. Some of the followers of Christian Wolff seemed determined to extract from every element of natural order an advantage for humankind and a trace of the Creator's wisdom. "*Vermessen*" (literally, *mismeasured*), Kant wrote,

> is a good word, full of meaning. A judgment in which we forget to estimate the extent of our powers may at times sound very humble, and yet it makes vast claims and is very presumptuous. Of that sort are most of those judgments in which we purport to exalt divine wisdom by attributing the works of creation and preservation to divine intentions that are actually meant to give credit to the wisdom of the very person who does the subtle reasoning. (ibid., p. 383n)

Judgments of relative purposiveness, the teleological constraint of one natural phenomenon for the sake of another, were, to Kant, invariably flawed. If, like Linnaeus, one proved that some natural phenomenon or the whole of nature existed for the sake of something else, there was no guarantee that something was itself an end of nature.

> We can easily see . . . that extrinsic purposiveness (a thing's being beneficial to others) can be regarded as an extrinsic natural purpose only under the condition that the existence of what it benefits proximately or remotely is a purpose of nature in its own right. This, however, we can never tell by merely examining nature; and hence it follows that, although relative purposiveness points hypothetically to natural purposes, it does not justify any absolute teleological judgment. (pp. 368–69)

Because naturalists treated organisms in terms of natural purposes, they also sought a principle beyond blind efficient causes for the economy of nature. To the degree that this led to the use of finality as a signpost pointing to a transcendental source for nature, they entangled themselves in difficulties. But the same judgment, if limited to the immanent structure of nature, had a certain justification. Insight into natural mechanism was limited, and judgments of relative purposiveness might allow a naturalist "to discover many further laws of nature that would otherwise remain hidden from us" (p. 398). In the limited sphere of scientific investigation, finality was only a mode of ordering and unifying natural phenomena, alternative to the mode of mechanism.

Of course, a proper scientific investigation must begin with mechanical principles. "Reason is tremendously concerned not to abandon the mechanism nature employs in its products and not to pass over it in explaining them, since without mechanism we cannot gain insight into the nature of things" (p. 410). Science had the right to approach the products and events of nature as mechanisms as far as this was possible. Inevitably, investigation would arrive at a point where the possibility of organized beings could not be accounted for in mechanical terms. At that point, science had the right to invoke causality acting in terms of purpose.

This was not a step to be taken lightly. To open a treatise on natural order by invoking the act of creation in the manner of Linnaeus did not advance science one whit. Science had no insight into transcendental acts and principles. All the same, the use of the principle of purpose in relation to the products of nature could not be discarded. The principle could not explain the source of living forms, but as a heuristic device, it might guide investigation where mechanical principles were inadequate.

Basic to Kant's conception of an appropriate use of teleological judgments in relation to natural order was the physical unity of the reproductive force active in a determinate manifold of animals. Adopting Buffon's genealogical method, Kant traced the unity and lawlike character of natural groups to their origin in a common historical lineage.[17] When joined to the idea of a prototype, a first design upon which natural groups seemed to have been conceived, the genealogical method would result in a physical system for the understanding. All of this required teleological assumptions. The naturalist had to assume some original organization with a predisposition to preserve and modify itself and with the capacity to produce others of its kind. With these assumptions in hand, a naturalist could tackle the problems of variation, relations among groups, and the history of nature with some hope of discovering the system of their production.

Variations triggered by external causes, but passed along through reproduction, could be judged only as the development of a predisposition originally present in the species. The naturalist had to assume that nothing was taken up into the generative force that did not already belong to the species, undeveloped original predispositions.[18] Similarly, a comparative anatomist sorted through the "vast creation of organized beings in nature" in the expectation of discovering a natural system.

> So many genera of animals share a certain common schema on which not only their bone structure but also the arrangement of their other parts seem to be based: the basic outline is admirably simple but yet able to produce this great diversity of species, by shortening some parts and lengthening others, by the involution of some and the evolution of others. Despite all the variety among these forms, they seem to have been produced according to a common archetype, and this analogy among them reinforces our suspicion that they are actually akin, produced by a common original mother. (ibid., p. 418)

Finally, the archeologist of nature was free to trace the genesis of the great family of living beings from nature's earliest revolutions.

> He can suppose that the womb of mother earth as it first emerged, like a huge animal, from its chaotic state gave birth to creatures whose form displayed less finality, and that these again bore others, which adapted themselves more perfectly to their native surroundings and their relations to each other, until this womb, becoming rigid and ossified, restricted its birth to definite species incapable of further modification. . . . Yet, for all that, he is obliged eventually to attribute to this universal mother an organization suitably constituted with a view to all these forms of life, for unless he does so, the possibility of the final form of the products of the animal and plant kingdoms is unthinkable. But when he does attribute all this to nature, he has only pushed the explanation a stage further back. He cannot pretend to have made the genesis of the kingdoms intelligible independently of the condition of final causes. (ibid., pp. 419–20)

Conclusion

With Kant's echo of Buffon, these studies in the sciences of living form have come full circle. By examining some of the ways in which inquiry began to free itself from the hold of older received ideas, these studies have shown how the sciences of living form succeeded in defining not a solution to the problem of a physical method but the signification of the problem.

In 1749 Buffon argued that science must avoid as far as possible having recourse to causes outside nature. His method rested upon observation and experiment, recognized the limitations of experience, and conceived investigation as a gradually unfolding process, increasing in clarity and precision as knowledge of the facts improved. Neither Haller nor Linnaeus accepted Buffon's formulation of method, of course, but they too were committed to the study of nature in a proper physical context. All three men conceived of nature as a self-regulating system, unfolding by means of physical processes in forms that could be studied and understood in physical terms.

The scientific achievements of the triumvirs rested squarely upon this commitment. Buffon's genealogical method redefined species in terms of reproduction and located the repetitions and variations of living forms along a temporal axis; his reform of generation replaced the immutability of preformation with formation over time and argued that development, like any other physical process, was subject to other physical conditions. Haller's experiments on irritable parts demonstrated that irritability operates mechanically and, under certain conditions, automatically; his observations of incubated eggs established organization as the central issue in the science of generation. Linnaeus's *Species plantarum* rested upon the inspection of no less than 5,900 species classed in 1,098 genera; his dissertation on the economy of nature analyzed, on the basis of observed effects, the physical principles that preserved the system of nature.

But the science of the triumvirs and their generation cannot be understood solely on the basis of a commitment to the study of living nature as a unified and self-contained physical system. Although they proposed to investigate nature on her own terms, they continued to presuppose a primordial formative act, whose purposiveness their empirical concepts also reflected. Buffon's species concept allowed for variation among living kinds and for the heredity of acquired characters; simultaneously, he thought of the natural species as fixed elements of an immutable natural order, without which science would have been impossible.

The forces that shaped matter in Haller's theory of evolution were also products of God's hand; once conferred, they could be taken for granted, and science could study the process of development in purely mechanical terms. Linnaeus could not have undertaken the classification of nature if he had not assumed that nature would disclose an arrangement whose source seemed to him obvious. In these and many other ways, naturalists and physiologists continued to draw upon the idea of an absolute teleological understanding at the heart of living nature.

Apart from these contradictory results, the scientific program of the earlier enlightenment was nonetheless important and influential. If the task of science was to secure the intelligibility of living nature, this was best done by turning away from final causes and the wonders of creation and concentrating instead on natural forms and physical processes. This part of the program involved generational and institutional quarrels with those who insisted that the autonomy of living nature was being subtracted from the divine patrimony. The program was not limited, however, to the substitution of nature for the creator. It also exalted the human capacity to know nature, if not absolutely, at least within the limits of observation and experiment. Living forms and processes, subject to all manner of physical conditions, required the study of all kinds of external and internal agents. An immense field of investigation opened up to the sciences of living form.

The response of the next generation was to divide the field and investigate it piecemeal. The immediate result was a multiplication of problems and interests. Between 1760 and 1790, this generation of scientists established the relevance of the new disciplines of comparative anatomy, embryology, ethnography, and geographical history to the older sciences of living form.

The unifying element in these divergent tendencies was a concern for exact observational methods appropriate to the specific and distinct subject of living nature. The concern is obvious in the generation of 1760's criticism of their elders—in their preference for accurate, well-coordinated information, in their tenacious, step-by-step pursuit of observation, and in their refusal to claim more than their methods could establish. "I do not say," Caspar Friedrich Wolff wrote in 1768, "that formation takes place through the concourse of particles, through the mode of fermentation, through mechanical causes and effects—but only that it does take place." And in saying so, Wolff spoke for many other members of his scientific generation.

The concern for exactitude and precision is clearest in the selection and treatment of problems. Joseph Gaertner, for instance, identified fructification as the functional system essential to any natural method for plants, then spent twenty years pursuing the comparative anatomy of fruit and seed in all its unimaginable detail. Joseph Koelreuter settled on the problem of plant hybridization and, in an exhaustive experimental series, not only limited the possibility of novelty by means of specific crosses but also showed that organic change was the product of a narrowly defined combinatory of characters. Peter Simon Pallas refuted the concept of degeneration by demonstrating that the external and internal causes adduced by Buffon had not changed the form, proportion, or inner structure of any species to a degree that would sustain a thesis of transformation. Eberhard Zimmermann organized the scattered information on distribution and showed that no previous conjecture had accounted for the existing stations and habitations of plants and animals.

It would certainly be going too far to claim that the generation of 1760 succeeded where their elders had failed, eliminating all traces of the plan of creation or nature's intentions from their physical methods. But there was a significant change: almost imperceptibly, older received ideas were emptied of their explanatory force and employed as heuristic devices in the process of investigation. In a sense, nature was trivialized and lost much of her metaphysical space. The generation of 1760 strove for what was clear and unequivocal, arranged in exact formal and causal relations that reflected the order and intelligibility of physical nature.

Joseph Gaertner certainly believed in the objective existence of a system of nature and projected a natural method to reflect that structure as accurately as possible. But he concerned himself with the physical identity and arrangement of the natural orders and rarely spoke of the source or purpose of this arrangement, even in programmatic statements. Joseph Koelreuter made no secret of his belief that the natural species had issued from the hand of the Creator, but he used his experiments to emphasize the quantifiable facts of plant sexuality and reproduction. Caspar Friedrich Wolff construed the relation between nourishment and structure as an essential force uniting teleological and mechanical principles, but he insisted that the essential force was a designation for effects observed a posteriori, a descriptive tool that in no way explained the cause of generation.

These are some of the ways in which the increasing sophistication of physical methods gradually emptied received ideas of ontological content and put them to work in the limited sphere of physical investigation.

And it is this process, the gradual reduction of metaphysical realities to the status of heuristic devices, that makes Kant's critique of teleological judgment so exemplary a statement of the development of physical method in the later eighteenth century. Kant followed this development attentively over a period of forty years, and in 1790 he asked whether it was possible and rational, at one and the same time, to deal with living forms according to mechanical principles and also to regard them from the vantage of purpose and organize and arrange them accordingly. As forms of knowledge, he argued, the two principles did not exclude one another. Mechanical causality dealt with order in change, while the concepts of purpose had to do with the structure of living forms; to acknowledge this structure was not to exclude mechanical causes. Organisms could not be completely understood in terms of mechanical principles, and there was some justification for introducing teleological considerations to compensate for the shortcomings of mechanical methods.

These considerations remained purely regulative; they were, to Kant, only "one principle more for submitting appearances to rules, where the laws of mechanical causality did not suffice." As long as naturalists and physiologists did not insist that nature herself was intelligent, or posit a being outside nature as her architect, they were even justified in speaking of nature's wisdom and foresight, since in the context of scientific inquiry these terms helped keep in view the special rules that science must follow in investigating living forms.

From its inception, the abstract program of secondary causes, physical processes, and intelligible principles was used to oppose an older science of final causes, immutable forms, and creationist timetables. Simultaneously, the physical program was a step in the direction of a new conception of nature, a nature who pursued the continuities of her own development by means of her own forces. As naturalists and physiologists established the relative autonomy of nature in the creation and arrangement of living forms, they began, quite consistently, to view nature as a whole engaged in her own development. But this insight is far more obvious to us than it was to scientists engaged in realizing a physical program.

The introduction of history and the time that succeeded itself into the atemporal sciences of natural history and physiology is a complicated story, far more complicated than any of the research yet devoted to the subject has allowed. For the most part, that story also lies outside the time limits of these studies. Before 1790, a conservative estimate of the significance of historicist constructions for the sciences of living form seems justified, although this estimate has proved very difficult to understand or accept. Twentieth-century historians of science know that historical thinking acquired a new impetus and direction at the end of the eighteenth century. They know that scientific theory in the disciplines of cosmology, geology, paleontology, and geography was already powerfully affected by historicist constructions. And they know that the introduction of historical thinking into the hitherto atemporal sciences of natural history and physiology led to an instauration of the life sciences in the next century. What more natural, then, than to generalize from these tendencies and to trace the entry of time and history into the life sciences to the later eighteenth century?

In many areas of scientific investigation, one can find theoretical and speculative statements that indicate a new interest in viewing living forms in terms of their history. In the upper reaches of the system of nature, naturalists were increasingly aware of profound structural concordances among the members of classes and orders of plants and animals. At the level of specific diversity, ideas of physical descent were increasingly important. In the study of distribution, historicist theses were introduced to explain patterns of discontinuity. And in the science of generation, the theory of epigenesis seemed to account for problems of heredity and environment without a solution in the doctrine of preformation.

But the connection between these interests and the assumptions and procedures of naturalists and physiologists remained tenuous. Naturalists continued to represent structural concordances in terms of atemporal affinities and analogies that left the source of unity undetermined. Genealogical methods did not change the morphological criteria with which naturalists established species. Historicist theses in geographical history were separated from the purely synchronic coordination of stations and habitations. And in the science of generation, mechanist conceptions of formation simply substituted an improbable arrangement of matter for an equally incomprehensible preexisting germ. There is an intellectual distance, noted by nearly every specialist who has dealt with the subject, between the theoretical and methodological formulas and the traditional procedures of these older sciences. And of the two, it is the procedures, not the formulas, that best indicate the position of these sciences toward the historicist reconstruction of the disciplines.

The great obstacles to the implementation of historicizing tendencies in the sciences of living form were the primacy of class concepts in natural history and the dominance of mechanical medicine in physiology. As legacies from earlier eras, class concepts and the anatomical deduction of function were laden with metaphysical assumptions. These assumptions were not directly subject to analysis and criticism, however, because they constituted the basis of university instruction and had become a part of the mind-set of professional naturalists and physiologists.

In the sciences of natural history, all systems and methods rested upon a traditional doctrine of concept formation. The fundamental categorical relation of an object to its properties was central to this doctrine, while relational determinations were considered only when they could be transformed into properties of an object or objects. This epistemological base controlled description and nomenclature, the composition of books, and the constitution of herbaria, cabinets, and gardens. Built up out of abstract affinities and analogies, the sciences of botany and zoology did not question whether formal likenesses merely indicated or actually constituted the relation between forms, and these sciences were for the most part indifferent to the physical source of the relation.

In physiology, Albrecht von Haller and his followers were notable exponents of mechanical medicine during the later eighteenth century. More or less in the tradition of Galen's *De usu partium*, they continued to conceive organs as artificial, premeditated constructions and to deduce organic function from structure. Their physiology, as the theory of generation shows very clearly, was essentially an atemporal science, tacitly based on the anatomy of adult forms, preferring invisible and undemonstrable coexistence to progressive genesis, and full of constancies and regularities inaccessible to scientific investigation.

To be fair, some of the elements that led to the modification or reversal of these views were already integral parts of these same sciences. Haller's experiments on the irritable and sensible parts accredited belief in the existence of physiological properties without evident connection to anatomical structures. Wolff's alternative science demonstrated the existence of novelties, transformations, and disappearances during the course of organic formation, unaccounted for in any theory of preformation. In natural history, the reluctant recognition of internal agents as the main source of specific diversity threatened that cornerstone of orthodox science, specific constancy.

Between 1760 and 1790, the sciences of natural history and physiology took an increasing interest in the problems of change and novelty, in common group forms, in physical descent, in the factors of environment and heredity, and in the autonomous dynamisms of living nature. But all of this did not add up to the history of nature projected by Immanuel Kant in the *Kritik der Urtheilskraft*—not in professional science, at any rate.

After 1790 the situation of the sciences of living form changed rapidly. The separate problems dealt with in these studies—the common structures of groups of plants and animals, the causes of diversity, the laws of distribution, and the mechanisms of formation—began to be integrated.

The forms and appearances of life were read over against the conditions, causes, and laws of life. In short order, the results began to approximate Kant's archeology of nature.

Simultaneously, the union of physicomechanical and teleological principles, which Immanuel Kant so generously attributed to the works of Johann Friedrich Blumenbach, reversed the traditional perspective on the process of formation. Mechanists had tried to show how an external force arranged relatively unformed materials in a coordinated whole. Blumenbach, on the other hand, began with the living form as a coordinated whole and adopted teleological principles for effects that he could not explain mechanically. He traced the regularity of those effects to an internal and particular finality, a formative force proper to living form. With Kant's powerful endorsement, Blumenbach went on to supplement the formative force with other vital forces and, in short order, had created a new and influential vitalist physiology.

Whatever their merits as science, transformism and vitalism strengthened the autonomy of the life sciences, although the justification for this autonomy continued to be a matter of dispute. Immanuel Kant ascribed it to the peculiar methods with which scientific inquiry took cognizance of living nature, and in doing so, he summed up the period in which appropriate physical methods were the center of scientific concerns. Other, younger heads, less willing to deny themselves what Kant had forbidden, saw the autonomy of the life sciences as that of a state within a state, based upon qualities peculiar to living matter and alien to physicomechanical principles.

A Note on the Notes

The titles of some articles have been shortened by cropping irrelevant phrases. Book titles always begin with the original word or phrase. Where necessary they are shortened internally by ellipsis. Subtitles and the final phrases of lengthy titles are usually omitted.

Arabic numerals are used both for the volume number of journals and for sections of multivolume works. The following abbreviations are used:

AAW	*Abhandlungen der Königlichen Akademie der Wissenschaften in Berlin*
Acta	*Acta Academiae Scientarum Imperialis Petropolitanae*
ADB	*Allgemeine deutsche Biographie*. 56 vols. Berlin: Duncker and Humblot, 1967
Annales	*Annales du Muséum National d'Histoire Naturelle par les Professeurs de cet Etablissement*
Comm	*Commentationes Societates Regiae Scientarum Gottingensis*
DSB	*Dictionary of Scientific Biography*. 18 vols. New York: Charles Scribner's Sons, 1970–90
GA	*Göttingsche gelehrte Anzeigen*
GS	*Kant's Gesammelte Schriften*, herausgegeben von der Königlich Preussischen Akademie der Wissenschaften
Handlingar	*Kungliga Svenska Vetenskapsakademiens Handlingar*
HAS	*Histoire de l'Académie Royale des Sciences et Belles Lettres*, à Berlin
HS	*The British Journal for the History of Science*
JHB	*Journal of the History of Biology*
JHI	*Journal of the History of Ideas*
MAS	*Mémoires de l'Académie Royale des Sciences et Belles Lettres*, à Berlin
NC	*Novi Commentarii Academiae Scientarum Imperialis Petropolitanae*
PT	The Royal Society of London. *Philosophical Transactions*
Proceedings	The Linnaen Society of London. *Proceedings*
RHS	*Revue d'Histoire des Sciences*

SA	*Sudhoffs Archiv für Geschichte der Medizin und der Naturwissenschaften*
Skrifter	*Skrifter af Carl von Linné*, utgifna af Kungliga Svenska Vetenskapsakademien, 4 vols. Upsala: Almqvist and Wiksell, 1905–08
SLSÅ	*Svenska Linné-Sällskapets Årsskrift*
Werke	*Naturwissenschaftliche Schriften*, *Goethes Werke*, 2. Abtheilung. 13 vols. Weimar: Böhlau, 1890–96
ZA	*Zoologische Annalen, Zeitschrift für Geschichte der Zoologie*

Notes

Introduction

1. Ernst Mayr, *The Growth of Biological Thought: Diversity, Evolution, and Inheritance* (Cambridge, Mass.: Harvard University Press, Belknap Press, 1982), p. 387.
2. Ernst Troeltsch, *Aufsätze zur Geistesgeschichte und Religionssoziologie, Gesammelte Schriften*, 4 vols. (Tübingen: J. C. B. Mohr, Paul Siebeck, 1924–25), 4: 532–33.
3. Bentley Glass, Owsei Temkin, and W. L. Strauss, Jr., eds., *The Forerunners of Darwin, 1745–1859* (Baltimore: Johns Hopkins University Press, 1959).
4. Friedrich Meinecke, *Die Entstehung des Historismus, Werke*, ed. Carl Hinrichs (Munich: Oldenbourg, 1965), p. 9.
5. These include standard works by Maurice Caullery, *A History of Biology*, trans. James Walling (New York: Walker, 1966); Erik Nordenskiöld, *The History of Biology* (New York: Tudor, 1928); Philip C. Ritterbush, *Overtures to Biology* (New Haven: Yale University Press, 1964). See also Joseph A. Caron, "'Biology' in the life sciences: a historiographical contribution," *HS* 26 (1988): 223–68; Jacques Roger, "Réflexions sur l'histoire de la biologie," *RHS* 8 (1964): 25–40.
6. Michel Foucault, *Les mots et les choses, une archéologie des sciences humains* (Paris: Gallimard, 1966), p. 139.
7. Joseph Schiller, *La notion d'organisation dans l'histoire de biologie*, préface par Pierre-P. Grassé (Paris: Maloine, 1978), p. 4. See also Schiller's *Physiologie et classification, relations historiques* (Paris: Maloine, 1980).
8. Claire Salomon-Bayet, *L'institution de la science et l'expérience du vivant: Méthode et expérience à l'Académie Royale des Sciences 1666–1793* (Paris: Flammarion, 1978).
9. Caron, "Biology."
10. Birgitte Hoppe, "Le concept de biologie chez G. R. Treviranus," in *Colloque international Lamarck*, ed. Joseph Schiller (Paris: Blanchard, 1971), pp. 199–237. See also Walter Baron, "Gedanken über den ursprünglichen Sinn der Ausdrücke Botanik, Zoologie, und Biologie," *SA*, Beiheft 7 (1966): 1–10.
11. Wolf Lepenies, *Das Ende der Naturgeschichte: Wandel kultureller Selbstverständlichkeiten in den Wissenschaft des 18. und 19. Jahrhunderts* (Munich: Hanser, 1976), pp. 28, 29.

12. Timothy Lenoir, "The Göttingen school and the development of transcendental Naturphilosophie in the romantic era," *Studies in the History of Biology* 5 (1981): 190; also see Lenoir, *The Strategy of Life: Teleology and Mechanics in Nineteenth-Century German Biology* (Dordrecht: Reidel, 1982), p. 1.

13. Gottfried Reinhold Treviranus, *Biologie, oder Philosophie der lebenden Natur für Naturforscher und Ærzte*, 6 vols. (Göttingen: Röwer, 1802–22), 1:1.

14. Jacques Roger, "The living world," in *The Ferment of Knowledge*, ed. G. S. Rousseau and Roy Porter (Cambridge: Cambridge University Press, 1980), p. 262. See, however, René Taton, ed., *Enseignement et diffusion des sciences en France au XVIII*[e] *siècle* (Paris: Hermann, 1964); and Salomon-Bayet, *L'institution de la science*.

15. Roger, "The living world," p. 259.

16. John Heilbron, "Introductory essay," in *The Quantifying Spirit in the Eighteenth Century*, ed. Tore Frängsmyr, John Heilbron, and Robin Rider (Berkeley: University of California Press, 1990), p. 23.

17. Roger, "The living world," p. 260.

18. Heilbron, "Introductory essay," pp. 22–23.

19. John Heilbron, *Elements of Early Modern Physics* (Berkeley: University of California Press, 1982), p. ix.

I Two Generations

1. Two introductory sections in Johann Friedrich Blumenbach's *Handbuch der Naturgeschichte*, 1st ed. (Göttingen: Dieterich, 1779), "Von Naturalien überhaupt," and "Von den organisirten Körpern überhaupt," are comparisons of the fixist definition of species with Buffon's concept of degeneration. The results vary in the many editions. See also Peter Simon Pallas, "Mémoire sur la variation des animaux," *Acta*, 1780 (1784): 70–80, and E. A. W. Zimmermann, *Geographischen Geschichte des Menschen und der allgemeinen verbreiteten vierfüssigen Thiere*, 3 vols. (Leipzig: Weygand, 1778–83), 3:196–98. By the mid-1780s the comparison of Linnaeus and Buffon was an established topic in natural history and lasted at least until Georges Cuvier's *Histoire des sciences naturelles depuis leur origine jusqu'à nos jours chez tous les peuples connus*, completée par M. Magdeleine de Saint-Agy, pt. 3, vols. 4–5 (Paris: Fortin, Masson, et Cie, 1843–45), 4:110–83.

2. Georges-Louis Le Clerc, Comte de Buffon, *Histoire naturelle générale et particulière, avec la description du Cabinet du Roi*, 15 vols. (Paris: L'imprimerie Royale, 1749–67).

3. Pallas, "Variation des animaux," pp. 74, 70.

4. See Ernst Cassirer, *The Philosophy of the Enlightenment*, trans. Fritz C. Koelln and James P. Pettegrove (Princeton: Princeton University Press, 1951), pp. 77–80.

5. Phillip Sloan, "The Buffon-Linnaeus controversy," *Isis* 67, no. 238 (1976): 356–75.

6. René-Antoine Ferchault de Réaumur and Bernard de Jussieu, colleagues of Buffon, were two particularly appropriate candidates. Buffon's attack on Linnaeus had been meditated as early as 1745. See Jean Piveteau, *Buffon, 1788–1988* (Paris: Imprimerie Nationale, 1988), pp. 162–63.

7. Buffon, *Oeuvres philosophiques de Buffon*, vol. I of Corpus Générale des Philosophes Français, XLI, texte établi et présenté par Jean Piveteau (Paris: Presses Universitaires de France, 1954), pp. 11–13, 18–19. For similar criticisms, see Lo-

renz Heister, *Systema plantarum generale* (Helmstadt: Weygand, 1748), and Jacob Theodor Klein, *Summa dubiorum* (Dantzig: Schreiber, 1743).

8. Buffon, *Oeuvres*, p. 10.

9. Phillip Sloan, "Buffon, German biology, and the historical interpretation of biological species," *HS* 12 (1979): 112–21.

10. Buffon, *Oeuvres*, pp. 23–26.

11. "Je crois, que le peu de connaissance que M. de Buffon a des auteurs systématiques est ce qui l'a empêché de faire attention à la première et principale untilité des méthodes." Lamoignon-Malesherbes, cited in Piveteau, *Buffon, 1788–1988*, p. 195.

12. Buffon, *Allgemeine Historie der Natur nach allen besondern Theilen abgehandelt*, ed. and trans. Abraham Kästner, 8 vols. (Hamburg, Leipzig: Grund und Holle, 1750–72).

13. Albrecht von Haller, Foreword to ibid., 1:x.

14. Kästner, in ibid., 1:36–37n; 2:78n.

15. See Sloan, "Buffon, German biology," p. 148.

16. Albrecht von Haller, *Primae lineae physiologiae in usum praelectionum academicarum* (Göttingen: Vandenhoek, 1747), p. 478.

17. Birger Strandell, "Linnés tackskrivelse till franska vetenskapsakademien 1763," *SLSÅ*, Årg. 1972–74 (1978): 153–56.

18. Carl Linnaeus, *Vita Caroli Linnaei: Carl von Linnés självbiografier*, ed. Elis Malmeström and Arvid Hjalmar Uggla (Stockholm: Almqvist & Wiksell, 1957), p. 133.

19. All quotations from Th. M. Fries, *Linné, lefnadsteckning*, 2 vols. (Stockholm: Fahlcrantz, 1903), 1:271–73.

20. After 1733 Buffon spent the winter months in Paris and the other eight months at Montbard. Between 1749 and 1768 he completed his fifteen-volume analysis of the quadrupeds. Besides his scientific work, Buffon was an estate owner, an academician, and an official of the French state.

During the fifties Haller published his experiments on the sensible and irritable parts of the body. He then began his observations on incubated eggs that led to his memoires on the formation of the heart and bone. In 1757 he began publishing the eight volumes of the *Elementa physiologiae corporis humani*. Haller kept up an incredible round as an administrator, controversialist, correspondent, reviewer of books, and supplier of texts.

In 1751 Linnaeus published the *Philosophia botanica*, the bible of the Linnaean method. In 1753 he published the *Species plantarum*, in which he first implemented his binomial nomenclature on a grand scale. In 1758 Linnaeus issued a tenth edition of the *Systema naturae*. This list omits all mention of Linnaeus's activities as a professor, academician, patron of scientific travel, and administrator.

21. For brief biographies of Caspar Friedrich Wolff (1734–94), see *ADB*, 44:41–43, and *DSB* 15:524–26. Wolff's letter is reprinted in Julius Schuster, "Der Streit um die Erkenntnis des organischen Werdens im Lichte der Briefe C. F. Wolffs an A. Von Haller," *SA* 34 (1941): 202. An English translation of the latter is in Shirley A. Roe, *Matter, Life, and Generation: Eighteenth-Century Embryology and the Haller-Wolff Debate* (Cambridge: Cambridge University Press, 1981), pp. 158–59.

22. Albrecht von Haller, "Review of C. F. Wolff, *Theoria Generationis*," *GA* 143

(1760): 1126–31. Haller's review is reprinted in Caspar Friedrich Wolff, *Theorie von der Generation in zwo Abhandlungen erklärt und bewiesen* (Berlin: Birnstiel, 1764), pp. 136–40.

23. Wolff, *Theorie von der Generation*, pp. 31–33, 74–97.

24. Linnaeus, *Politia naturae*, in *Skrifter* (1906), pp. 155–76.

25. Linnaeus, "Nitraria planta obscura explicata a Carolo Linnaeo," *NC* 7, 1758–59 (1761): 315–20.

26. Linnaeus, "Nya bevis för sexualitet hos växterna," in *Skrifter* (1908).

27. Joseph Gottlieb Koelreuter, *Procès-verbaux, Académie Impériale des Sciences* (1744–70) (St. Petersburg, 1897), 2:453. For a biography of Koelreuter, see Robert Olby, *Late Eighteenth Century European Scientists* (Oxford: Pergamon, 1966), pp. 33–65, and *DSB* 7:440–42.

28. Joseph Gottlieb Koelreuter, *Vorläufige Nachricht von einigen das Geschlecht der Pflanzen betreffenden Versuchen und Beobachtungen nebst Fortsetzugen 1, 2, und 3 (1761–1766)*, ed. W. Pfeffer (Leipzig: Engelmann, 1893).

29. Carl Otto von Sydow, "Linnaeus and Gmelin," *SLSÅ*, Årg. 1978 (1978): 212–22.

30. For biographies of Peter Simon Pallas (1741–1811), see *ADB* 25: 81–98, and *DSB* 10:283–85.

31. Peter Simon Pallas, *Elenchus zoophytorum, sistens generum adumbrationes generaliores et specierum cognitarum* (The Hague: van Cleef, 1766). I cite, however, a German translation: Pallas, *Characteristik der Tierpflanzen*, trans. Wilkens und Herbst (Nuremberg: Bauer, 1787).

32. Pallas, *Characteristik der Tierpflanzen*, pp. 46–47.

33. Pallas, "Variation des animaux."

34. Other names should be mentioned. The list is restricted to professors and academicians working in northern and central Europe whose work has been analyzed in the course of these studies: Johann Beckmann (1739–1811); Peter Jonas Bergius (1730–90); Torbern Bergman (1735–84); Nicholas Laurens Burman (1733–98); Johann Christian Polycarp Erxleben (1744–77); Johann Christian Fabricius (1745–1808); Johann Reinhold Forster (1729–98); Joseph Gaertner (1732–91); Johann Gottlieb Georgi (1729–1802); Paul Dietrich Giseke (1745–96); Samuel Gottlieb Gmelin (1744–74); Johann Anton Güldenstädt (1745–81); Johann Hermann (1738–1800); Nikolas Joseph von Jacquin (1727–1817); Otto Frederik Müller (1730–84); Johann Andreas Murray (1740–91); Georg Christian von Oeder (1728–91); Eduard Sandifort (1742–1814); Horace-Bénédict de Saussure (1740–99); Johann Christian Daniel Schreber (1739–1810); Anders Sparrman (1748–1820); Gottlieb Conrad Christian Storr (1749–1821); Cornelis Ascanius van Sypesteyn (1723–83); Carl Peter Thunberg (1743–1828); Johann Daniel Titius (1723–96); Heinrich August Wrisberg (1739–1808); Eberhard August Wilhelm Zimmermann (1743–1815); Johann Gottfried Zinn (1727–59).

35. Heilbron, "Introductory essay," pp. 1–2.

36. Wolff, *Theorie von der Generation*, p. 7.

37. J. C. D. Schreber, however, disavowed the use of "einen Maastab, oder andere Dinge, woraus man die Grösse der Thiere ohngefähr ermessen könnte." *Die Säugthiere nach der Natur mit Beschreibungen*, 7 vols. (Erlangen: Walther, 1775–1846), 1:iii.

38. Wolff, *Theorie von der Generation*, pp. 264–72.

39. Caspar Friedrich Wolff, *De formatione intestinorum*, Pt. 3, *NC* 13 (1768): 518–20. (Pts. 1 and 2 are in *NC* 12 (1766–67.)

40. Karl Ernst von Baer, *Über Entwickelungsgeschichte der Thiere*, 2 vols. (Königsberg: Bornträger, 1828–37), 2:121n.

41. For a biography of Gaertner, see Joseph Deleuze, "Notice sur la vie et les ouvrages de Gaertner," *Annales* 1 (1802): 207–33; *DSB* 5:216–17.

42. Cuvier, *Histoire des sciences naturelles* 5:288.

43. Heilbron, "Introductory essay," p. 23. For a brief survey of these institutions and their functions, see Heilbron, *Elements of Early Modern Physics*, pp. 107–25. For more detail, see James E. McClellan III, *Science Reorganized: Scientific Societies in the Eighteenth Century* (New York: Columbia University Press, 1985), and Robin Rider, "Bibliographical Afterword," in Frängsmyr, Heilbron, and Rider, eds., *The Quantifying Spirit*, pp. 387–91.

44. Linnaeus, "Tal om märkwärdigheter uti insecterna," in *Tre tal*, ed. Arvid Hjalmar Uggla (Stockholm: Geber, 1954).

45. Cited in Ragnar Granit, *Utur Stubbotan Rot: Essäer till 200-årsminnet av Carl von Linnés död* utgivna af Kungl. Vetenskapsakademien (Stockholm: Norstedt & Söners, 1978), p. 7.

46. Cited in Sten Lindroth, "The two faces of Linnaeus," in *Linnaeus: The Man and His Work*, ed. Tore Frängsmyr (Berkeley: University of California Press, 1983), p. 61.

47. Cited in Roe, *Matter, Life, and Generation*, p. 83.

48. Johann Friedrich Blumenbach, *Anfangsgründe der Physiologie*, trans. Joseph Eyerel (Vienna: Wappler, 1795), pp. 242–43.

49. Cited in Piveteau, *Buffon, 1788–1988*, p. 164.

50. Buffon, *Oeuvres*, p. 243.

51. Pallas, "Variation des animaux," p. 72.

52. Albrecht von Haller, *Réflexions sur le système de la génération de M. de Buffon* (Geneva: Barillot et Fils, 1751), p. 54.

53. Zimmermann, *Geographische Geschichte* 3:192. For a biography of Zimmermann, see *ADB* 45:256–58.

II The System of Nature

1. Linnaeus, *Politia naturae*, p. 157. Cf. Camille Limoges, *C. Linné: L'équilibre de la nature*, trans. Bernard Jasmin (Paris: Vrin, 1972), pp. 7–22.

2. Christian von Wolff, *Philosophia prima sive Ontologia* (Frankfurt, Leipzig: Renger, 1730), p. 155. See also Christian von Wolff, *Vernünftige Gedancken von Gott, der Welt, und der Seele des Menschen, und allen Dingen überhaupt*, 3d ed. (Halle: Renger, 1724), p. 337: "Eine Machine ist ein zusammengesetzte Werck, dessen Bewegungen in der Art der Zusammensetzung gegründet sind. Die Welt ist gleichfalls ein zusammengesetztes Ding, dessen Veränderungen in der Art der Zusammensetzung gegründet sind. Und demnach ist die Welt eine Machine." On order and truth, see Wolff, *Vernünftige Gedancken*, pp. 337f; on the *nexus rerum*, see Christian von Wolff, *Cosmologia generalis* (Frankfurt: Renger, 1737), pp. 9ff. See also Tore Frängsmyr, *Wolffianismens genombrott i Uppsala* (Uppsala: Almqvist and Wiksell, 1972), pp. 21–36.

3. Gottfried Wilhelm Leibniz, *New Essays on Human Understanding*, ed. Peter

Remnant and Jonathan Bennett (Cambridge: Cambridge University Press, 1981), p. 56. Cf. Giorgio Tonelli, "The law of continuity in the eighteenth century," *Studies on Voltaire and the Eighteenth Century* 27 (1963): 1619–38.

4. August Thienemann, "Die Stufenfolge der Dinge, der Versuch eines natürlichen Systems der Naturkörper aus dem achtzehnten Jahrhundert," *ZA* 3 (1910): 235–46. Cf. Lorin Anderson, *Charles Bonnet and the Order of the Known* (Dordrecht: Reidel, 1982), pp. 34–58.

5. Charles Bonnet, *Betrachtung über die Natur*, trans. Johann Daniel Titius (Leipzig: Junius, 1772). Cf. Thienemann, "Stufenfolge," pp. 243–44.

6. Jean-Baptiste-René Robinet, *De la nature*, 4 vols. (Amsterdam: van Harrevelt, 1761–66) 4:11–12.

7. Jacques Roger, *Les sciences de la vie dans la pensée françise au xviiie siècle: La génération des animaux de Descartes à l'Encyclopédie*, 2d ed. (Paris: Colin, 1971), p. 648.

8. Buffon, *Oeuvres*, p. 24.

9. Henri Daudin, *Cuvier et Lamarck: Les classes zoologiques et l'idée de série animale*, 2 vols. (Paris: Alcan, 1926), 2:239–75.

10. Frans Stafleu, *Linnaeus and the Linnaeans: The Spreading of Their Ideas in Systematic Botany, 1735–89* (Utrecht: Oosthoek's Uitgeversmaatschappij N. V., 1971).

11. Gunnar Eriksson, "The botanical success of Linnaeus: The aspect of organization and publicity," *SLSÅ*, Årg. 1978 (1978): 57–66.

12. Mary Winsor, "The development of Linnaean insect classification," *Taxon* 25, no. 1 (1976): 57–67.

13. J. C. P. Erxleben, Foreword to *Anfangsgründe der Naturlehre*, 2d ed. (Göttingen: Dieterich, 1777), unpaginated.

14. The work of Christian Gottlieb Ludwig (1709–73) offers an introduction to these concerns. Ludwig accompanied Hebenstreit to North Africa, where he observed the fertilization of palms and pistacia. Upon his return, Ludwig defended a thesis on the sexuality of plants, *De sexu plantarum* (1737). That same year, he compared the systems of Rivinus and Linnaeus in *Definitiones plantarum* and worked out the characters of some genera more carefully. In *Observationes in methodum Linnaei* (1739), he noted that the division of plant sexes was not always essential, and he questioned the grouping of some Linnaean genera.

These concerns were constant preoccupations of the Linnaean party until the end of the century. Other members of the party include Johann Gesner, Johann Ernst Stieff, Carl August von Bergen, and Phil. Conr. Fabricius. Johann Gottlieb Gleditsch's pollination of a palm should be mentioned here: "Essai d'une fécondation artificielle fait sur l'espece de palmier qu'on nomme, *Palma Dactylifera folio flabelliformi*," *HAS*, 1749 (1751): 103–08.

15. The obscurities involved in applying the sexual theory are clear in work devoted to mushrooms. G. F. Märcklin, an apothecary in Weislock, regarded mushrooms as sports of nature, not to be divided into genus and species. N. J. von Necker argued that they were products of crystallization. There was no agreement about sexual organs. J. G. Gleditsch, "Conjecture sur l'usage des corps diaphanes de Michelius dans les champignons à lames," and "Experience concernant le génération des champignons," *HAS*, 1748–49 (1750–51): 60–66; 26–32, argued that the diaphanous bodies of Michelius, surrounded by female organs, served in place

of seminal powder. Joseph Gottlieb Koelreuter, *Das entdeckte Geheimnis der Cryptogamie* (Karlsruhe: Macklot, 1777), argued that the volva was a male organ. See also Johannes Hedwig, *Theoria generationis et fructificationis plantarum cryptogamicarum* (St. Petersburg: Typis Academicae imper. scientiarum, 1787). Much the same story was repeated in work devoted to the other cryptogams.

16. J. M. Hulth, *Bibliographia Linnaeana: Matériaux pour servir à une bibliographie Linnéenne* (Uppsala: Almqvist and Wiksell, 1907), pp. 2–16, 23–27, 89–92.

17. J. C. P. Erxleben, *Anfangsgründe der Naturgeschichte* (Göttingen: Dieterich, 1773), p. 16.

18. Antoine-Laurent de Jussieu, *Genera plantarum secundum ordines naturales disposita* (Paris: Herissant, 1789), p. 72.

> Not only does the natural method convey a complete knowledge of characters and affinities, but it also, at the same time, indicates the virtues of plants, to the great advantage of medicine and the alimentary arts. . . . Plants of the same genus that are subsimilar in characters or organs, scarcely differ in their virtues: thus Salvias are all good for the head. . . . By this same further connection is gradually revealed the sequence of contiguous genera that are related in many organs, or in outstanding ones; those somewhat similar to one another in virtue are brought together, and observation confirms that this virtue is uniform in truly natural orders or in the genuine sections of orders. . . . Therefore, there exists the closest affinity between characters and virtues—so close, indeed, that the ones are harbingers, so to speak, of the others; in other words, from the known virtues of certain plants, we can almost certainly infer the unknown virtues of other, related ones. In this manner the natural method increases our knowledge of appropriate remedies and healthy food, and in this sedulous refinement of the medical and alimentary arts, it far surpasses the less fruitful, artificial systems.

19. Heister, *Systema plantarum generale.*

20. Augustin-Pyrame de Candolle, *Théorie élémentaire de la botanique*, 3d ed. (Paris: Roret, 1844), pp. 58–59.

21. Albrecht von Haller, *Historia stirpium indigenarum Helvetiae inchoata*, 2 vols. (Bern: Societatis Typographicae, 1768), 1:i–x.

22. Johann Gottlieb Gleditsch, "Système des plantes fondé sur la situation et liaison des étamines," *HAS*, 1749 (1751): 110.

23. Joseph Gaertner, *De fructibus et seminibus plantarum* (vol. 1, Stuttgart: Typis Academiae Carolinae, 1788; vol. 2, Tübingen: Typis Giulielmi Henrici Schrammis, 1791); the traditional term *carpology* is used for ease of identification.

24. Antoine-Laurent de Jussieu, "Examen de la Famille des Renoncules," *HAS*, 1773 (1777): 216.

25. Deleuze, "Vie de Gaertner."

26. Gaertner, *De Fructibus* 1:clxxv.

27. Among the first, Gaertner included "the consistency of the pericarp, the common receptacle, the albumen, the number of loculi, the valves, the seeds, the size and thickness of the embryo, the curves and folds of the cotyledons, and finally the absence or presence of the plumula, for all of these may differ in the species of one genus" (ibid., p. clxxvii).

28. Among characters deserving attention, Gaertner emphasized the situation of parts (pericarp, loculi, receptacle, and radicle), the form of the receptacle, the number of seeds, the arillus and the baccate covering, the straightness or curvature of the embryo, and differences in the forms of cotyledons. When these parts differed, Gaertner wrote, "genera ought to differ also, even when all other remaining parts are similar" (ibid.).

29. Deleuze, "Vie de Gaertner," p. 220.

30. "These genera, filled with few or many species at will, are to be called truly natural when they do not separate related species or disturb natural order." Jussieu, *Genera plantarum*, p. xxxvii.

31. "Some are changeable and varied, others more constant, and others supremely constant or essential: in plants to be compared, characters are not thus discussed indiscriminately but in the order mentioned" (ibid., p. xix).

32. Antoine-Laurent de Jussieu, "Premier Mémoire: Sur les caractères généraux de familles . . . rectifiées par les observations de Gaertner," *Annales* 5 (1804): 218.

33. Linnaeus, *Praelectiones in ordines naturales plantarum*, ed. Paul Dieterich Giseke (Hamburg: Hoffmann, 1792).

34. Cuvier, *Histoire des sciences naturelles* 5 : 298.

35. "The parts, properties, or traits of conformation that have the greatest number of these relations of incompatibility or coexistence with others, or in other words, which exert the most marked influence over the whole of the entity are those which are called important characters, and there are different degrees of them." A. J. Cain, "The post-Linnaean development of taxonomy," *Proceedings* 170, no. 3 (1959): 237. See also Walter Baron, "Methodologische Probleme der Begriffe Klassifikation und Systematik sowie Entwicklung und Entstehung in der Biologie," in *System und Klassifikation in Wissenschaft und Dokumentation*, ed. A. Diember (Meisenheim am Glan: Hain, 1968), pp. 15–31.

36. Jussieu, *Genera plantarum*, p. xix.

37. F. P. Cassel, *Lehrbuch der natürlichen Pflanzenordnung* (Frankfurt am Main: Andreäischen Buchhandlung, 1817), p. vi.

38. Lorenz Oken, *Lehrbuch der Naturphilosophie*, 2d ed. (Jena: Frommann, 1810), p. 220.

39. Cassel, *Lehrbuch*, p. 116.

40. "It is true," Cuvier continued, "that Daubenton and Camper had furnished some facts, that Pallas had indicated his views; but the ideas of these learned men still had not exercised upon their contemporaries the influence they deserved to have." Georges Cuvier, *Le règne animal distribué d'après son organisation*, 4 vols. (Paris: Déterville, 1817), 1 : vii.

41. Johann Friedrich Blumenbach, *Handbuch der Naturgeschichte*, 8th ed., 2 vols. (Göttingen: Dieterich, 1807), 1 : 61.

42. Cuvier, *Histoire des sciences naturelles*, 5 : 233.

43. Carl Friedrich Kielmeyer, *Ueber die Verhältnisse der organischen Kräfte unter einander in der Reihe der verschiedenen Organisationen . . . 1793* (Tübingen: Osiander, 1814).

44. Daudin, *Cuvier et Lamarck*, 1 : 86.

45. Cuvier, *Histoire des sciences naturelles*, 3 : 278–79. "Ces recherches furent re-

prises par Haller, par ces élèves et par ceux de Boerhaave et Albinus; car tous ces hommes, à quelques différences près, appartient à la même école."

46. For an introduction to the literature on microscopes, see Maurice Daumas, *Les instruments scientifiques aux XVIIe et XVIIIe siècles* (Paris: Presses Universitaires de France, 1953), pp. 211–39, 333–38. See also P. Harting, *Das Mikroskop: Theorie, Gebrauch, Geschichte* (Braunschweig: Vieweg und Sohn, 1859), and S. Bradbury, "The quality of the image produced by the compound microscope: 1700–1840," in *Historical Aspects of Microscopy*, ed. S. Bradbury and G. l'E. Turner (Cambridge: Heffer and Sons, 1967), pp. 151–73.

Also see Leonhard Euler's studies: "Règles générales pour la construction des télescopes et microscopes de quelque nombre de verres qu'ils soient composés"; "Détermination du champ apparent que découvrent tant les télescopes que les microscopes"; and "Recherches sur les microscopes à trois verres et les moyens de les perfectionner," *MAS* (1757, 1761, 1764), 12:283, 8:191, and 20:117.

Stagnation in optical properties was accompanied by an inventiveness in accessories. An apparatus for the study of small animals is illustrated in Daumas, *Les instruments scientifiques*, fig. 82.

47. J. Victor Carus, *Geschichte der Zoologie bis auf Joh. Müller und Charles Darwin* (Munich: Oldenbourg, 1872), pp. 479–80, mentions Jacob Theodor Klein and Johann Daniel Titius as classing animals according to habitat.

48. Linnaeus, *Systema naturae*, 10th ed., 2 vols. (Stockholm: Salvius, 1758–59), 1:11; Linnaeus, *Systema naturae*, 12th ed., 3 vols. (Stockholm: Salvius, 1766–68), 1:19.

49. Blumenbach, *Handbuch der Naturgeschichte*, 1st ed., p. 113.

50. Gottlieb Conrad Christian Storr, *Prodromus methodi mammalium* (Tübingen: Litteris Reissianis, 1780), pp. 13–16.

51. August Johann Georg Carl Batsch, *Versuch einer Anleitung zur Kenntnis und Geschichte der Thiere und Mineralien*, 2 vols. (Jena: Akademische Buchhandlung, 1788–89), 2:80–92.

52. Daudin, *Cuvier et Lamarck*, 1:114–15; A.-N. Duchesne, "Sur les rapports entre les êtres naturels," *Magasin Encyclopédique* 1, no. 6 (1795): 292.

53. Lamarck, "Discours d'ouverture du cours de zoologie," *Système des animaux sans vertèbres* (Paris: Déterville, 1801). See Daudin, *Cuvier et Lamarck*, 1:116–117n. Lamarck's use of the distinction dates back to his first course at the Muséum in the spring of 1794. Georges Cuvier, *Leçons d'anatomie comparée*, 2 vols. (Paris: Baudouin, 1800), 1:65.

54. Linnaeus, *Systema naturae*, 10th ed., 1:13.

55. Nils von Hofsten, "Linnés Djursystem," *SLSÅ*, Årg. 12 (1959): 32–33.

56. Linnaeus, *Systema naturae*, 10th ed., 1:608–40.

57. Blumenbach, *Handbuch der Naturgeschichte*, 1st ed., pp. 391–400, 401.

58. Daudin, *Cuvier et Lamarck*, 1:294.

59. Charles de Geer, *Mémoires pour servir à l'histoire des insectes*, 8 vols. (Stockholm: Hesselberg, 1752–78), 2:1–16; 7:1–7, 667–68.

60. Johann Christian Fabricius, *Philosophia entomologica* (Hamburg: Bohn, 1778), p. 94.

61. Georges Cuvier, "Mémoire sur la structure interne et externe et sur les affinités des animaux auxquelles on a donné le nom de Vers," *Decade philosophique* 1, no. 2 (1795): 433–49.

62. Linnaeus, *Systema naturae*, 1st ed. (Leyden: Haak, 1735), unpaginated (12th page).

63. Linnaeus, *Systema naturae*, 10th ed., 1:641–821. Cf. Hofsten, "Linnés djursystem," pp. 9ff, and Nils von Hofsten, "A system of 'double entries' in the zoological classification of Linnaeus," *Zoologiska bidrag från Uppsala* 35 (1963).

64. Johann Baptist Bohadsch, *De quibusdam animalibus marinis* (Dresden: Walther, 1761), pp. 2–3, 52–53, passim.

65. Blumenbach, *Handbuch der Naturgeschichte*, 1st ed., pp. 44–45.

66. Peter Simon Pallas, *Miscellanea zoologica quibus novae imprimis obscurae animalium species describuntur* (The Hague: van Cleef, 1766), pp. 72–74. In Italy, J. X. Poli accepted Linnaeus's system of double entries for cephalopods and gasteropods. See his *Introductio ad scientarium*, vol. 1 of *Testacea utriusque Sicili*, 3 vols. (Parma: Imprim. Royale, 1791–1827), 1:27–29.

67. Cuvier, "Mémoire sur la structure," pp. 394–96.

68. Daudin, *Cuvier et Lamarck*, 2:3–35.

69. Cain, "The post-Linnaean development of taxonomy," p. 237. See also A. J. Cain, "Deductive and inductive methods in post-Linnaean taxonomy," *Proceedings*, 170, no. 2 (1959): 185–217.

70. Vitaliano Donati, *Essai sur l'histoire naturelle de la Mer Adriatique* (The Hague: de Hondt, 1758), p. 20.

71. Johann Hermann, *Tabulae affinitatum animalium, olim academico specimine edita* (Strasbourg: Treuttel, 1783). See especially Hermann's discussion of affinity, "Affinitatem generatim," pp. 4–37.

72. Lorenz Oken, *Abrisses des Systems der Biologie* (Göttingen: Vandenhoek, 1805), pp. 230ff.

73. I cite the German translation, Pallas, *Characteristik der Tierpflanzen*, p. 46.

74. Giulio Barsanti, *Le immagini della natura: scale, mappe, alberi 1700–1800* (Florence: Olschki, 1988), pp. 81–99.

75. Peter F. Stevens, "Antoine-Laurent de Jussieu, 'Nature', and the Natural System" (Cambridge, Mass., December, 1991), unpublished manuscript.

76. Johann Friedrich Blumenbach, *Beyträge zur Naturgeschichte* (Göttingen: Dieterich, 1790), pp. 111, 112.

77. I use the German translation, Pallas, *Characteristik der Tierpflanzen*.

78. Daudin, *Cuvier et Lamarck*, 2:249–50.

III The Scale of Diversity

1. I cite Buffon, *Oeuvres*, p. 354.

2. Linnaeus, *Fundamenta botanica* (Amsterdam: Schouten, 1736), p. 18. See also Linnaeus, *Philosophia botanica* (Stockholm: Kiesewetter, 1751), p. 99; and his "Ratio operis," in *Genera plantarum*, 1st ed. (Leiden: Wishoff, 1737), unpaginated.

3. Johann Friedrich Blumenbach, *Über den Bildungstrieb und das Zeugungsgeschäfte* (Göttingen: Dieterich, 1781), pp. 57–58.

4. Linnaeus, *Flora lapponica*, ed. James Edward Smith (London: White and Sons, 1792); originally published by Schouten in Amsterdam in 1737). Alexander von Humboldt, *Essai sur la géographie des plantes accompagné d'un tableau physique des régions equinoxales* (Mexico: Institut panaméricain de géographie et d'histoire, 1955).

5. Kurt Sprengel, *Anleitung zur Kenntnis der Gewächse*, 2d ed., 2 vols. (Halle: Kummel, 1817–18), 1:368.
6. Linnaeus, *Genera plantarum*, 6th ed. (Stockholm: Salvius, 1764), unpaginated, following the index.
7. Pallas, "Variation des animaux," p. 75.
8. Ibid., pp. 74–75. Also see Sprengel, *Anleitung*, 1:375.
9. Candolle, *Théorie élémentaire de la botanique*, p. 167.
10. Pallas, "Variation des animaux," p. 73.
11. Candolle, *Théorie élémentaire*, p. 172.
12. Cited in Pallas, "Variation des animaux," p. 73n.
13. Candolle, *Théorie élémentaire*, p. 172.
14. Koelreuter, *Vorläufige Nachricht*; Karl Ludwig Willdenow, *Grundriss der Kräuterkunde zu Vorlesungen entworfen*, 5th ed. (Berlin: Haude und Spener, 1810); Sprengel, *Anleitung*.
15. Pallas, "Variation des animaux," p. 73.
16. Cited in Roe, *Matter, Life, and Generation*, pp. 128–29.
17. Koelreuter, *Vorläufige Nachricht*, p. 42.
18. Cited in Pallas, "Variation des animaux," p. 74n.
19. "The proposition was either to confirm or to deny, by means of new proofs and experiments as well as those already known, the doctrine of the sex of plants, preceded by a history and an account of all the parts of the plant that play some role in fructification and the formation of seed and fruit." Linnaeus, "Nya bevis för sexualitet hos växterna," *Skrifter* (1908), p. 129.
20. François Delaporte, *Nature's Second Kingdom*, trans. Arthur Goldhammer (Cambridge, Mass.; MIT Press, 1982), pp. 91–148.
21. Linnaeus, "Nya bevis för sexualitet hos växterna," p. 128.
22. Blumenbach, *Handbuch der Naturgeschichte*, 8th ed., 1:21.
23. Koelreuter, *Vorläufige Nachricht*, p. 29.
24. C. E. B. Bremekamp, "Linné's views on the hierarchy of taxonomic groups," *Acta Botanica Neerlandica* 2, no. 2 (1953): 246.
25. Robert Olby, *Late 18th-Century European Scientists* (Oxford: Pergamon, 1966), p. 53.
26. Koelreuter, *Vorläufige Nachricht*, pp. 44, 31.
27. Pallas, "Variation des animaux," pp. 82–83.
28. "I find hybrid plants more common than hybrid animals and rather many in number. I believe I have been allowed to open the door to one of nature's extensive chambers, although it is not opened without creaking." Linnaeus, *Bref och skrifvelser af och till Carl von Linné*, ed. Th. M. Fries and J. M. Hulth (Uppsala: Akademiska bokhandeln, 1907–22), 1, 4, p. 140.
29. Koelreuter, *Vorläufige Nachricht*, p. 132.
30. Mayr, *The Growth of Biological Thought*, p. 646.
31. *GA*, 102:889–91; *GA*, 36:281–84; and *GA* 26:207–8.
32. Blumenbach, *Über den Bildungstrieb*, p. 61.
33. Pallas, "Variation des animaux," p. 61.
34. Schreber followed the 12th edition of Linnaeus's *Systema naturae* in his *Die Säugethiere*. Schreber conceived this illustrated reference for a broad public and replaced Linnaeus's dry definitions with *ausführliche Beschreibungen* in the manner

of Buffon. Schreber was assiduous in quoting Buffon's opinions but only on the subject of his disposition of species.

J. C. P. Erxleben followed the 10th and 12th editions of the *Systema naturae* in his *Systema regni animalis* (Leipzig: Impensis Weygandis, 1777). He adopted Pallas's correction of Buffon's classification of gazelles and noted other Buffonian dispositions but did not mention the genealogical method. As a systematist, Erxleben alternated between essential characters and the entire habitus in determining groups. Besides Blumenbach's *Handbuch der Naturgeschichte* and Batsch's *Anleitung zur Kenntnis und Geschichte der Thiere und Mineralien*, Storr's *Prodromus methodi mammalium* should be mentioned. Storr pursued a "physiological" method, independent of questions of likeness or descent.

35. On this important point, Buffon and Daubenton agreed with Linnaeus and the Linnaeans.

36. Pallas, "Variation des animaux," pp. 76–77.

37. Zimmermann, *Geographische Geschichte*, 1:27.

38. Cuvier, *Histoire des sciences naturelles*, 4:170.

39. Pallas, "Variation des animaux," p. 78n.

40. Linnaeus, *Systema naturae*, 10th ed., 1:18–24.

41. See Gunnar Broberg, *Homo Sapiens L., Studier i Carl von Linnés naturuppfattning och människolära*, vol. 28, *Lychnos bibliotek* (Uppsala: Almqvist & Wiksell, 1975), pp. 164–95.

42. Buffon, *Oeuvres*, pp. 386–413.

43. Henry Home, Lord Kames, *Six Sketches on the History of Man*, 2 vols. (Edinburgh; London: Creech and Strahan; Cadell, 1774). Buffon's memoires appeared in German in 1772; Lord Kames's book was published in Leipzig in 1774. Cf. Johann Unold, *Die ethnologischen und anthropogeographischen Anschauungen bei I. Kant und J. Reinh. Forster* (Leipzig: Dietz'sche Hofbuchdr. Coburg, 1886), pp. 19f.

44. Immanuel Kant, "Von den verschiedenen Racen der Menschen," *GS*, sec. 1, vol. 2, p. 429.

45. Immanuel Kant, *Physische Geographie*, *GS*, sec. 1, vol. 9, p. 162.

46. Kant, "Verschiedenen Racen," pp. 429–30. See especially Phillip Sloan's discussion of this paper in "Buffon, German biology, and the historical interpretation of biological species," *HS* 12 (1979): 125–29.

47. Kant, "Verschiedenen Racen," p. 430.

48. Immanuel Kant, "Bestimmung des Begriffs einer Menschenrasse," *GS*, sec. 1, vol. 8, pp. 91–106.

49. Georg Forster, "Noch etwas über die Menschenrassen," *Werke*, vol. 8 (Berlin: Akademie Verlag, 1974).

50. Immanuel Kant, "Über den Gebrauch teleologischer Prinzipien in der Philosophie," *GS* 1, vol. 8, p. 180.

51. Blumenbach, *Handbuch der Naturgeschichte*, 8th ed., 1:3.

52. Linnaeus, *Fundamenta botanica*, p. 18.

53. Blumenbach, *Handbuch der Naturgeschichte*, 1:20.

54. Timothy Lenoir, "Kant, Blumenbach, and vital materialism in German biology," *Isis* 71, no. 256 (March 1980): 89.

55. Sloan, "Buffon, German biology"; Lenoir, "Vital materialism."

56. Blumenbach, *Anfangsgründe der Physiologie*; Johann Friedrich Blumenbach,

Über die naturlichen Verschiedenheiten in Menschengeschlechte, trans. Johann Gottl. Gruber (Leipzig: Breitkopf, 1798), pp. 259–61; Lenoir, "Vital materialism," p. 93.

57. Lenoir, "Vital materialism," pp. 90–91.

58. Blumenbach, *Beyträge zur Naturgeschichte*, pp. 22, 20n.

59. Blumenbach, *Beyträge zur Naturgeschichte*, p. 23.

60. Blumenbach, *Beyträge zur Naturgeschichte*, p. 61.

61. Christoph Girtanner, *Ueber das kantische Prinzip für die Naturgeschichte* (Göttingen: Vandenhoek und Ruprecht, 1796), pp. 3–4.

62. Lenoir, "Vital materialism," p. 97.

63. Girtanner, *Ueber das kantische Prinzip*, p. 2.

64. Sloan, "Buffon, German biology," pp. 141–42.

65. Johann Karl Wilhelm Illiger, *Versuch einer systematischen vollständigen Terminologie für die Thierreich und das Pflanzenreich* (Helmstädt: Fleckeisen, 1800), p. 140.

66. Johann Wolfgang von Goethe, "Der Verfasser theilt die Geschichte seiner botanischen Studien mit," *Werke*, sec. 2, vol. 6, p. 116.

67. Cited in Roe, *Matter, Life, and Generation*, p. 128.

IV Distribution of the Natural Kinds

1. The subject as a whole has attracted only two historians. Nils von Hofsten's paper, "Zur älteren Geschichte des Diskontinuitätsproblems," *ZA* 7 (1919): 197–353, is not widely known, but among those who have read it the paper is regarded as a classic discussion of the problem of discontinuity. With respect to the eighteenth century, however, Hofsten's use of the concept of *discontinuity* is anachronistic; no eighteenth-century naturalist unified his discussion of geographical distribution with this term. Moreover, Hofsten's focus on theory obliged him to overlook the organization given by earlier naturalists to the myriad of facts relating to geographical history. On the other hand, Janet Browne's *The Secular Ark: Studies in the History of Biogeography* (New Haven: Yale University Press, 1983) is selective and omits northern and central Europeans not available in English translation. In the case of both Hofsten and Browne, the tendency to anchor interpretations in nineteenth-century biogeography has simplified eighteenth-century problems.

2. Blumenbach, *Handbuch der Naturgeschichte*. 1st ed.; Pallas, "Variation des animaux," pp. 69–102; Zimmermann, *Geographische Geschichte*, 3:196–98.

3. Limoges, *L'équilibre de la nature*, pp. 7–22. See also Gunnar Broberg, *Carl von Linné Om jämvikten i naturen*, trans. Anders Piltz (Stockholm: Carmina, 1978), pp. 4–48.

4. Linnaeus, *Vita Caroli Linnaei*, p. 169. The Linnaean economy of nature was mercantilist, not liberal.

5. Linnaeus, *Politia naturae*, p. 172.

6. Limoges, *L'équilibre de la nature*, p. 12.

7. Linnaeus, "De telluris habitabilis incremento," in *Skrifter* (1906), p. 93. See also Browne, *The Secular Ark*, pp. 18–23.

8. See Tore Frängsmyr, "The Diminution of the Waters Controversy," in Frängsmyr, ed., *Linnaeus*, pp. 125–43.

9. Kurt Sprengel, *Von den Bau und der Natur der Gewächse* (Halle: Kummel, 1812), pp. 623–33. See also the memoire by B.-G.-E. de Lacépède, "Des hauteurs et des positions correspondantes des principales montagnes du globe, et de l'influence de ces hauteurs et des positions, sur les habitations des animaux," *Annales* 9 (1807): 303–18.

10. Buffon, *Histoire naturelle*, 9:96, 118.

11. Buffon, *Oeuvres*, pp. 168–76.

12. Cited in Cuvier, *Histoire des sciences naturelles*, 4:167–72.

13. See, for example, two papers by Sloan, "The Buffon-Linnaeus controversy," and "Buffon, German biology."

14. See also E. A. W. Zimmermann, *Specimen zoologiae geographicae quadrupedum domicilia et migrationes sistens* (Leiden: Theodorum Haak et Socios, 1777); and F. S. Bodenheimer, "Zimmermann's *Specimen* . . . a remarkable zoogeographical publication of the end of the 18th century," *Archives international d'histoire des sciences* 33 (1955): 351–57.

15. Zimmermann, *Geographische Geschichte*, 3:196.

16. For a bibliography, see Kurt Sprengel, "Botanischen Reisen und Kenntnis ausländischen Pflanzen," and "Untersuchung vaterländischen Pflanzen," in his *Geschichte der Botanik*, 2 ed., 2 vols. (Leipzig: Brockhaus, 1817, 1818), 2:321–24, 343–53.

17. For remarks on the Linnaean terminology adopted here, see Gareth Nelson, "From Candolle to Croizat: Comments on the history of biogeography," *JHB* 11 (1978): 280–81.

18. Linnaeus, *Flora lapponica*; Johann Georg Gmelin, *Flora sibirica, sive historia plantarum sibiriae*, 4 vols. (St. Petersburg: Typographiae Academiae Scientarum, 1742–69); Haller, *Historia stirpium.*

19. Candolle, *Théorie élémentaire de la botanique*, pp. 257–62.

20. Linnaeus, "Tankar om nyttiga växters planterande på Lappska fjällen," *Handlingar* 15 (1754): 183.

21. Linnaeus, *Flora lapponica*, p. xix. A standard work on Linnaean plant geography is Einar du Rietz, "Linné som fjällväxtgeograf," *SLSÅ*, Årg. 25 (1942): 35–57.

22. Linnaeus, *Flora lapponica*, p. xix.

23. Linnaeus took the idea from Joseph Pitton de Tournefort's *Relation d'un voyage du Levant*, where, Linnaeus noted, Tournefort "found at the foot of Mount Ararat those plants common in Armenia; a little further those he had seen in Italy; when he ascended somewhat higher, such vegetables as were common in Paris; the plants of Sweden possessed a more elevated region; but the highest tracts of the mountain, next to the summit, were occupied by the natives of the Swiss and Lapland Alps" (p. 93).

24. J. G. Zinn, "Von dem Ursprunge der Pflanzen," *Hamburgische Magasin* (1756): 344–46.

25. Gmelin, *Flora sibirica*, 1:cvi–cvii, cx. See Sydow, "Linnaeus and Gmelin," pp. 212–22.

26. *GA* (1747): 781.

27. Linnaeus, "Flora alpina," in *Amoenitates academicae* (Erlangen: Palm, 1788), 4:419.

28. Albrecht von Haller, *Enumeratio methodica stirpium Helvetiae indigenarum* (Göttingen: Vandenhoek, 1742).

29. Haller, *Historia stirpium*, 1:viii.

30. Peter Simon Pallas, *Reise durch verschiedene Provinzen des russischen Reichs*, 3 vols. (Graz: Akademische Druck und Verlagsanstalt, 1967), 1:311–21.

31. Dietmar Henze, Foreword to Pallas, *Reise*, 1:xv.

32. Peter Simon Pallas, *Flora rossica seu stirpium imperii Rossica per Europam et Asiam indigenarum* (St. Petersburg: Typographia Imperialii, 1784); Jean-Baptiste de Monet de Lamarck and Augustin-Pyrame de Candolle, *Flore française, ou descriptions succinctes de toutes les plantes qui croissent naturellement en France*, 3d ed., 2 vols. (Paris: Desray, 1815), 2:2. See the foldout map at the end of the volume, with Candolle's explication, pp. v–xii.

33. Ramond de Carbonnière, *Observations faites dans les Pyrenées pour servir de suite à des observations sur les Alpes* (Paris: Belin, 1789), p. 331.

34. Lamarck and Candolle, *Flore française*, 2:vii.

35. Georg Forster, "Gentiana Saxosa, en obekant art ifrån Nya Zeeland," *Handlingar* 38 (1777):183.

36. Johann Reinhold Forster, *Beschreibungen der Gattungen von Pflanzen auf einer Reise nach der Inseln der Süd-See* (Stuttgart: Mantler, 1779; the Latin original, *Characteres*, was published in 1776).

37. Johann Reinhold Forster, *Observations Made During a Voyage Round the World on Physical Geography, Natural History, and Ethic Philosophy* (London: Robinson, 1778), p. 134.

38. Georg Forster, "Fasciculus plantarum magellicanarum," *Comm* 9 (1789): 17–18.

39. Hofsten, "Diskontinuitätsproblems," p. 252.

40. Willdenow, *Grundriss der Kräuterkunde*, p. 479.

41. Treviranus, *Biologie*, 2:131–36.

42. See J. L. Larson, "Vital forces: Regulative principles or constitutive agents?" *Isis* 70, no. 252 (1979):235–49.

43. Humboldt, *Essai*, p. 22. See W. T. Stearn, *Humboldt, Bonpland, Kunth, and Tropical American Botany* (Stuttgart: Cramer, 1968), pp. 121–28.

44. Humboldt, *Essai*, p. 20.

45. Zimmermann, *Geographische Geschichte*, 1:27–28.

46. Cited in Hofsten, "Diskontinuitätsproblems," p. 263.

47. Cited in Stearn, *Humboldt, Bonpland*, pp. 125–26.

48. Humboldt, *Essai*, pp. 19–22.

49. Augustin-Pyrame de Candolle, "Géographie botanique," *Dictionnaire des sciences naturelles* (Strasbourg: Levrault et le Normant, 1820), 18:412–19.

50. Zimmermann, *Geographische Geschichte*, 1:12.

51. "Zimmermann gab dieser Untersuchung eine solche Ausdehnung, wendete so grossen Fleiss und so genaue Kritik darauf und entwickelte die daraus herzuleitenden Folgen mit so vielen Scharfsinn, dass ein späteren Bearbeiter desselben Gegenstandes wenig mehr als das Verdienst sich erwerben kann, welches ihm ein Zeitraum von dreissig für die Naturkunde sehr ergiebigen Jahren gewährt." J. C. W. Illiger, "Ueberblick der Säugthiere nach ihrer Vertheilung über die Welttheile," *AAW*, 1804–11 (1816):39.

52. Fabricius, *Philosophia entomologica*, pp. 154–58.
53. Pierre-André Latreille, *Mémoires sur divers sujets de l'histoire naturelle* (Paris: Deterville, 1819), pp. 166–95.
54. J. R. Forster, *Enchiridion historiae naturali inserviens* (Halle: Hammerde und Schwetschke, 1788), pp. 29, 77–78, 138–39.
55. J. C. W. Illiger, "Tabellarische Uebersicht der Vertheilung der Vögel über die Erde," *AAW*, 1812–15 (1816):222.
56. Illiger, "Säugthiere," pp. 65, 85, 92, 98, 117, passim; Illiger, "Vögel," pp. 222–23.
57. Zimmermann, *Geographische Geschichte*, 1:239.
58. K. A. Rudolphi, *Beyträge zur Anthropologie und allgemeine Naturgeschichte* (Berlin: Haude und Spener, 1812), p. 143.
59. H. Fischer, "Mutmassliche Gedanken von dem Ursprung der Amerikaner," *Neue nordische Beyträge* 3 (1782):321.
60. Pallas, *Reise*, 3:290.
61. Pallas, "Observations sur la formation des montagnes et les changements arrivés au globe, particulièrement à l'égard de l'Empire de Russie," *Acta*, 1777, *pars prior* (1778):63n.
62. Pallas, *Zoographica rosso-asiatica*, 3 vols. (St. Petersburg: Academiae Scientarum, 1811–31), 1:115.
63. Fischer, "Mutmassliche Gedanken," p. 322.
64. See Christoph Meiners's review of Forster, *Voyage Round the World*, *GA*, Zugabe (1778):154.
65. J. R. Forster, "Historia aptenodytae," *Comm* 3(1778):121–48.
66. Treviranus, *Biologie*, 2:6–26.
67. Illiger, "Säugthiere," pp. 39–40; Illiger, "Vögel," pp. 222–23.
68. Rudolphi, *Beyträge*, pp. 130, 140–44.
69. Georg Forster, *Werke* (Berlin; Akademie, 1974), p. 151.
70. Hofsten, "Diskontinuitätsproblems," p. 266.
71. Candolle, "Géographie botanique," pp. 361–62.
72. Zimmermann, *Geographische Geschichte*, 2:196.

V The Mechanism of Formation

1. Cited in Howard B. Adelmann, *Marcello Malpighi and the Evolution of Embryology*, 5 vols. (Ithaca: Cornell University Press, 1966), 2:893–94. On the meaning of *evolutio*, see Peter J. Bowler, "The changing meaning of evolution," *JHI* 36 (1975):95–114; Georges Canguilhem, "Du développement à l'évolution au XIXe siècle," *Thales* 2 (1960):3–63; Stephen Jay Gould, *Ontogeny and Phylogeny* (Cambridge, Mass.: Harvard University Press, Belknap Press, 1977), pp. 28–32.
2. Buffon, *Oeuvres*, p. 282.
3. See, for example, Adelmann, *Malpighi*; Peter J. Bowler, "Preformation and Pre-existence in the Seventeenth Century: A Brief Analysis," *JHB* 4 (1971): 221–44; Elizabeth Gasking, *Investigations into Generation, 1651–1828* (Baltimore: Johns Hopkins University Press, 1967); Robert Herrlinger, "Einführung," in Caspar Friedrich Wolff, *Theorie von der Generation, Theoria Generationis* (Hildesheim: Olms, 1966), pp. 5–28; Jane Oppenheimer, *Essays in the History of Embryology and*

Biology (Cambridge, Mass.: MIT Press, 1967); Roe, *Matter, Life, and Generation*; Roger, *Les sciences de la vie*; Patrick Tort, "L'ordre du corps," in Maupertuis, *Vénus physique suivi de la lettre sur le progrès des sciences* (Paris: Montaigne, 1980), pp. 9–53.

4. This account follows Roger, *Les sciences de la vie*, pp. 542–58. See also Roger, *Buffon, un philosophe au jardin du roi* (Paris: Fayard, 1989), pp. 166–84.

5. Virginia P. Dawson, *Nature's Enigma: The Problem of the Polyp in the Letters of Bonnet, Trembley, and Réaumur*, Memoirs of the American Philosophical Society 174 (Philadelphia: American Philosophical Society, 1987); Aram Vartanian, "Trembley's polyp, La Mettrie, and eighteenth-century French materialism," *JHI* 11 (1950): 259–86.

6. Maupertuis, *Vénus physique*, p. 110.

7. Roger, *Les sciences de la vie*, p. 752.

8. Maupertuis, *Vénus physique*, p. 108.

9. René-Antoine Ferchault de Réaumur, *The Art of Hatching and Bringing up Domestic Fowls of All Kinds, at Any Time of the Year*, trans. Abraham Trembley (London: Davis, Millar, and Nourse, 1750), p. 463.

10. Pierre-Louis Moreau de Maupertuis, *Systéme de la nature*, vol. 2 of *Oeuvres*, 4 vols. (Lyon: Bruyset, 1756), pp. 146–47.

11. Buffon, *Oeuvres*, p. 248.

12. Ibid., p. 252. Roger gives a detailed account of Buffon's observations in *Buffon*, pp. 185–206.

13. Buffon, *Oeuvres*, p. 243.

14. Roger, *Les sciences de la vie*, p. 754.

15. Maupertuis, *Vénus physique*, p. 124.

16. To mention only the best known, Julien Offray de la Mettrie, *Histoire naturelle de l'ame* (The Hague: Néaulme, 1745); Mettrie, *L'Homme-machine*, ed. Aram Vartanian (Princeton: Princeton University Press, 1960); John Turberville Needham, "A summary of some late observations upon the generation, composition, and decomposition of animal and vegetable substance," *PT* 45 (1748): 615–66; Needham, *Nouvelles observations microscopiques* (Paris: Ganeau, 1750); Buffon, *Histoire naturelle*; Louis-Jean-Marie Daubenton, "Botanique," in *Encyclopédie, ou Dictionnaire raisonné des sciences, des arts, et des métiers* (1751), 2:340–45; Daubenton, "Exposition des distributions méthodiques des animaux quadrupèdes," "Description du cheval," in Buffon, *Histoire naturelle*, 4:142–68, 258–367; Denis Diderot, "Animal," and "De l'Interpretation de la Nature," in Diderot, *Oeuvres complètes* (Paris: Hermann, 1976), 5:381–400; 9:26–111.

17. Cited in Roger, *Les sciences de la vie*, p. 704.

18. Haller, *Réflexions*; and Haller, Foreword to Buffon, *Allgemeine Historie der Natur*, vol. 2.

19. Buffon, *Oeuvres*, p. 238; Diderot, "Animal," p. 400.

20. Haller, *Réflexions*, p. 24.

21. Haller, Foreword to Buffon, *Allgemeine Historie der Natur*, 2:xv.

22. Haller, *Réflexions*, p. 60.

23. Otto Sontag, ed., *The Correspondence Between Albrecht von Haller and Charles Bonnet* (Bern: Huber, 1983), p. 53.

24. Albrecht von Haller, *Sur la formation du coeur dans le poulet, sur l'oeil, sur la structure du jaune . . .*, 2 vols. (Lausanne: Bousquet, 1758), 1:186.

25. Roe, *Matter, Life, and Generation*, pp. 158–59.

26. Albrecht von Haller, *Elementa physiologiae corporis humani*, 8 vols. (Lausanne; Bern: Bousquet, d'Arnay, Grasset; Societas Typographica, 1757–66), 8:143.

27. Caspar Friedrich Wolff, *Theoria Generationis* (Halle: Litteris Hendelianis, 1759); also see Schuster, "Der Streit um die Erkenntnis," pp. 196–218; and Roe, *Matter, Life, and Generation*, pp. 158–59.

28. Caspar Friedrich Wolff, *Von der eigenthümlichen und wesentlichen Kraft der vegetabilischen sowohl als auch der animalischen Substanz* (St. Petersburg: Kayserlichen Akademie der Wissenschaften, 1789), pp. 48–50.

29. Wolff, *Theoria Generationis*, p. 13

30. Wolff, *Von der eigenthümlichen und wesentlichen Kraft*, p. 49.

31. Wolff, *Theoria Generationis*, p. 73.

32. Canguilhem, "Du développement à l'évolution," p. 6.

33. Cited in A. E. Gaissinowitsch, "Notizen von C. F. Wolff über die Bemerkungen der Opponenten zu seiner Dissertation," *Wiss. Zeitschrift der Friedrich-Schiller-Universität Jena; Math.—Naturwiss. Reihe*, 3/4 (1956–57): 121.

34. Haller, "Review of C. F. Wolff, *Theoria Generationis*," here cited from the reprint in Wolff's *Theorie von der Generation*, p. 137.

35. Gaissinowitsch, "Notizen von C. F. Wolff," p. 121.

36. Alfred Kirchhoff, *Zweiter Jahresbericht Luisenstädtische Gewerbeschule in Berlin* (Berlin: Schade, 1867), pp. 6–8. Kirchhoff gives some insight into medical instruction and the medical faculty at the Berlin Collegium.

37. Wolff, *Theorie von der Generation*, pp. 2–14.

38. Schuster, "Streit um die Erkenntnis," p. 208; Roe, *Matter, Life, and Generation*, p. 163.

39. Wolff, *Theorie von der Generation*, p. 106.

40. See Roe, *Matter, Life, and Generation*, pp. 161–62.

41. Wolff, *Theorie von der Generation*, pp. 267–68.

42. Cited in Roe, *Matter, Life, and Generation*, p. 40.

43. Wolff, *De formatione intestinorum* pts. 1 and 2, pp. 403–507; pt. 3, pp. 478–530.

44. Cf. August Johann Rösel von Rosenhof, *Historia naturalis ranarum nostratium oder: die natürliche Historie der Frösche hiesigen Landes* (Nürnberg: Fleischmann, 1758).

45. Canguilhem, "Du développement à l'évolution," p. 8.

46. Caspar Friedrich Wolff, *C. F. Wolff über die Bildung des Darmkanals im bebrüteten Hühnchen*, trans. Johann Friedrich Meckel (Halle: Renger, 1812), pp. 138–39n.

47. Haller's review of part 3 of Wolff's *De formatione intestinorum* is in *GA*, Zugabe (1771):415.

48. Wolff, *Von der eigenthümlichen und wesentlichen Kraft*, pp. ii–iii.

49. Blumenbach, *Erste Abhandlung über die Nutritionskraft* (St. Petersburg: Kayserlichen Academie der Wissenschaften, 1789).

50. Wolff, *Von der eigenthümlichen und wesentlichen Kraft*, p. 3.

51. Kant, *Briefe*, *GS*, sec. 2, vol. 2, p. 176.

52. Blumenbach, *Anfangsgründe der Physiologie*, p. 463.

53. Treviranus, *Biologie*, 1:21; C. C. E. Schmid, *Physiologie, philosophisch bearbeitet* (Jena: Akad., 1798–1801), 1:273, 371; J. B. Erhard, "Über die Möglichkeit der

Heilkunst," *Röschlaubs Magazin* 1 (1799): 1; Alexander von Humboldt, *Aphorismen aus der chemischen Physiologie der Pflanzen*, trans. Gotthelf Fischer (Leipzig: Voss, 1794).

54. Humboldt, *Aphorismen*, p. 3.

55. Kielmeyer, *Ueber die Verhältnisse der organischen Kräfte*, pp. 40–41.

56. Johann Friedrich Meckel, *C. F. Wolff über die Bildung des Darmkanals* (Halle: Renger, 1812), p. 4.

57. Canguilhem, "Du développement à l'évolution," pp. 9–10.

VI Kant and the Critique of Teleological Judgment

1. Immanuel Kant, *Kritik der Urtheilskraft, GS*, sec. 1, vol. 5. Cf. Wolfgang Bartuschat, "Neuere Arbeiten zu Kants Kritik der Urteilskraft," *Philosophische Rundschau* 18 (1972): 161–89; Bartuschat, *Zum systematischen Ort von Kants Kritik der Urteilskraft*, Philosophische Abhandlungen 43 (Frankfurt am Main: Klostermann, 1972); Paul Bommersheim, "Der Begriff der organischen Selbstregulation in Kants Kritik der Urteilskraft," *Kant-Studien* 23 (1919); Bommersheim, "Der vierfache Sinn der inneren Zweckmässigkeit in Kants Philosophie des Organischen," *Kant-Studien* 32 (1927): 290–309; Klaus Düsing, *Die Teleologie in Kants Weltbegriff*, Kantstudien Ergänzungshefte 96 (Bonn: Bouvier, 1968); Reinhard Löw, *Philosophie des Lebendigen, den Begriff des organischen bei Kant, sein Grund und seine Aktualität* (Frankfurt am Main: Suhrkamp, 1980); Konrad Marc-Wogau, *Vier Studien zu Kants Kritik der Urteilskraft*, vol. 2, Uppsala Universitets Årsskrift 1938 (Uppsala: Lundequistska, 1938); Giorgio Tonelli, "Von den verschiedenen Bedeutungen des Wortes 'Zweckmässigkeit' in der *Kritik der Urteilskraft*," *Kant-Studien* 49 (1957–58): 154–66.

2. "So ist doch von empirischen Gesetzen eine *so unendliche Mannigfaltigkeit* und *eine so grosse Heterogeneität* der Formen der Natur, die zur besondern Erfahrung gehören würden, möglich, dass der Begriff von einem System nach diesen (empirischen) Gesetzen dem Verstande ganz fremd sein muss, und weder die Möglichkeit, noch weniger aber die Notwendigkeit eines solchen Ganzen begriffen werden kann." Immanuel Kant, *Erste Einleitung in die Kritik der Urtheilskraft, GS*, sec. 3, vol. 7, p. 203.

3. Immanuel Kant, *Briefwechsel, 1789–1794, GS*, sec. 2, vol. 2, p. 426.

4. Kant, *Erste Einleitung*, pp. 213–14.

5. "Nun ist klar, dass die reflektierende Urtheilskraft es ihrer Natur nach nicht unternehmen könne, die ganze Natur nach ihren empirischen Verschiedenheiten zu klassifizieren, wenn sie nicht voraussetzt, die Natur spezifiziere selbst ihre transzendentalen Gesetze nach irgendeinen Prinzip." Ibid., p. 215.

6. Kant, *Kritik der Urtheilskraft*, p. 186.

7. The passage draws upon Blumenbach's definition of the *Bildungstrieb*. Cf. Blumenbach, *Über den Bildungstrieb*, pp. 12–13.

8. "Es ist doch etwas Anderes, ob ich sage: die Erzeugung gewisser Dinge der Natur, oder auch der gesammten Natur ist nur durch eine Ursache, die sich nach Absichten zum Handeln bestimmt, möglich; oder ich kann nach der eigenthümlichen Beschaffenheit meiner Erkenntnisvermögen über die Möglichkeit jener Dinge und ihre Erzeugung nicht anders urtheilen, als wenn ich mir zu dieser eine Ursache, die nach Absichten wirkt, mithin ein Wesen denke, welches nach der

Analogie mit der Causalität eines Verstandes productive ist" Kant, *Kritik der Urtheilskraft*, pp. 397–98.

9. Blumenbach, *Anfangsgründe der Physiologie*, p. 463.

10. Kant, *Erste Einleitung*, p. 236; Kant, *Kritik der Urtheilskraft*, p. 360.

11. Kant, *Kritik der Urtheilskraft*, pp. 389, 375.

12. Kant, *Erste Einleitung*, pp. 236–37.

13. Kant, *Kritik der Urtheilskraft*, p. 384.

14. Kant, *Erste Einleitung*, p. 235.

15. Kant, *Kritik der Urtheilskraft*, pp. 374, 388–89.

16. Ibid., p. 427.

17. The idea has deep roots in Kantian criticism as well. We understand a particular manifold when we reconceive it in terms of a principle specifying itself. If we apply this schema to living forms, we make the unity of natural groups comprehensible through reproductive modifications of an original stock.

18. "Denn wenn man von diesem Princip abgeht, so kann man mit Sicherheit nicht wissen, ob nicht mehrere Stücke der jetzt an einer Spezies anzutreffenden Form eben so zufälligen Ursprungs sein mögen; und das Princip der Teleologie: in einem organisirten Wesen nichts von dem, was sich in der Fortpflanzung desselben erhält, als unzweckmässig zu beurtheilen, müsste dadurch in der Anwendung sehr unzuverlässig werden und lediglich für den Urstamm (den wir nicht mehr kennen) gültig sein" (ibid., p. 420).

Bibliographical Afterword

Older histories of the individual sciences provide a solid point of departure for detailed information about specific concepts, practices, and tasks of natural history and physiology in the eighteenth century. Kurt Sprengel's *Geschichte der Botanik*, 2d ed., 2 vols. (Leipzig: Brockhaus, 1817, 1818), is particularly thorough for the German-speaking world. An equivalent work in zoology is J. Victor Carus, *Geschichte der Zoologie bis auf Joh. Müller und Charles Darwin* (Munich: Oldenbourg, 1872). For medicine Kurt Sprengel again is very detailed, *Versuch einer pragmatischen Geschichte der Arzneykunde*, 6 vols. (Halle, Vienna: B. Eble, 1823–26; 1837, 1840).

Comprehensive treatments of the life sciences in the eighteenth century by two later participants deserve more attention than they have yet received. Georges Cuvier concluded his career with a survey of the highways and byways of the life sciences. Since his days as a student at the Carlsschule, he had followed research with passionate engagement and intelligence, all of which he distilled one last time in the two volumes devoted to the later eighteenth century, *Histoire des sciences naturelles depuis leur origine jusqu'à nos jours chez tous les peuples connus*, complétée par M. Magdeleine de Saint-Agy, pt. 3, vols. 4, 5 (Paris: Fortin, Masson, et Cie, 1843–45). By way of contrast, Gottfried Reinhold Treviranus's ambitious attempt to weld natural history and physiology into a single philosophy of living nature is as instructive for its failures as for its successes: *Biologie, oder Philosophie der lebenden Natur für Naturforscher und Ærzte*, 6 vols. (Göttingen: Röwer, 1802–22). The views of both Cuvier and Treviranus are, of course, very personal, to be read, as Haller said of Buffon, "mit einer philosophischen Achtsamkeit."

Among twentieth century histories of the eighteenth-century sciences of living form, two works are mandatory reading. Henri Daudin's study of zoologists and zoology at the Muséum in Paris continues to serve as a quarry for research on systematics and comparative anatomy: *Cuvier et Lamarck: Les classes zoologiques et l'idée de série animale*, 2 vols. (Paris: Alcan, 1926). Jacques Roger's *Les sciences de la vie dans la pensée française au xviiie siècle: La génération des animaux de Descartes à l'Encyclopédie*, 2d ed. (Paris: Colin, 1971), is far more than a history of the science of generation. Roger's treatment of Maupertuis, Needham, Buffon, and Diderot has no match. A third, posthumous edition and an English translation have been announced.

Bibliographical Afterword

Although sociological and institutional studies of the life sciences in eighteenth-century Europe are in short supply, there are copious materials for these studies, textbooks, correspondences, and learned transactions for all of the major universities and academies of northern and central Europe. The scientific centers—Leiden, Göttingen, Tübingen, Leipzig, Halle, Berlin, St. Petersburg, Copenhagen, Stockholm, and Uppsala—require separate preliminary studies. A possible model for these studies can be found in the work of four generations of scholars at the Institut för idé- och lärdoms-historia at Uppsala. In particular, Sten Lindroth's histories of the Swedish Academy of Science, of Uppsala University, and of Swedish learning, pay close attention to the professional and institutional contexts of Swedish natural history and medicine.

The collection, arrangement, and preservation of natural objects was a matter of passionate concern to eighteenth-century scientists and their patrons. Just to list the accessions of one of the great explorers (Peter Simon Pallas, for example) or the contributions of one of the great patrons (Joseph Banks, for example) would require a large study. And at every major scientific center, specialists sorted through the latest finds to provide an eager scientific public with novelties. At Göttingen, for example, a professor of medicine, Johann Andreas Murray, annually published Latin descriptions of rare new plants, along with life-sized engravings, in the *Commentationes*.

The literature devoted to scientific hardware—microscopes, especially—is not exactly small. This too was a matter of passionate concern to the scientific community in the eighteenth century. The optical problems of color dispersion and spheric aberrations attracted some of the keenest minds of the century—Samuel Klingenstierna, Leonhard Euler, John Doland, and Alexis-Claude Clairaut, to name the most important. In the history of the life sciences, these concerns often boil down to the practical question of what could be seen, and how clearly. A more pertinent question might be, How did scientific preconceptions affect the interpretation of what was seen?

Because theory, principle, and worldview will continue for the foreseeable future to serve as the center of the history of the life sciences, the works of the great system builders, Buffon, Haller, and Linnaeus, provide a point of departure for the study of science in the later eighteenth century. Throughout their long careers, these men were very productive. Mastery of the views of any one of them does not come easily or quickly; no one can pretend to be equally knowledgeable about the work of all three. What follows are some suggestions for approaching these very different productions.

Buffon scholarship has concentrated upon the analysis of the quadrupeds, *Histoire naturelle générale et particulière, avec la description du Cabinet du Roi*, 15 vols. (Paris: Imprimerie Royale, 1749–67), but knowledge of this work is uneven. Interest focuses on the "Premier discours," the "Théorie de la terre," the "Histoire générale des animaux," and the memoires on "L'Asne," "Nomenclature des singes," "Dégeneration des animaux," and so on. There are relatively few references to the later works on birds (9 vols., 1770–83) or minerals (5 vols., 1783–88). Even the supplements to the original *Histoire naturelle* are not well known. One exception to the scholarly indifference to Buffon's later production is the *Époques de la nature*, which originally appeared as vol. 5 of the *Supplément* in 1778.

A critical edition of this important work was published by Jacques Roger and should be consulted in that form, *Les Époques de la nature*, Mémoires du Muséum National d'histoire naturelle, nouvelle série C, vol. 10 (Paris: Editions du Muséum, 1962). Since few libraries can boast a complete set of Buffon, Jean Piveteau's edition of the *Oeuvres* is a practical and reliable alternative, *Oeuvres philosophiques de Buffon*, vol. 41 of *Corpus Générale des Philosophes Français* texte établie et présenté par Jean Piveteau (Paris: Presses Universitaires de France, 1954). Besides the important discourses and memoires, Piveteau generously included the text of *Des Époques de la nature*, a selection from early and late work, and an important Buffon bibliography by Mme. E. Genet-Varcin and Jacques Roger. The German translation of the *Histoire naturelle* offers a running commentary on Buffon's unorthodox ideas: *Allgemeine Historie der Natur nach allen besondern Theilen abgehandelt*, 8 vols. (Hamburg; Leipzig: Grund; Holle, 1750–72). The studies by Piveteau, Roe, Roger, and Sloan cited repeatedly in the notes to these studies have determined my own interpretation of Buffon.

Of the triumvirs, Albrecht von Haller remains the most difficult to study. His learned activity was so great and so various, and was broadcast so widely, that to this day we do not have a complete picture of this great figure. Specialization in limited areas of Haller's production is the tried-and-true solution to this problem. These studies deal with Haller's theories of generation and organic distribution. In both cases, close readings of his books have been supplemented with comments in reviews and correspondence. This has meant the construction of careful chronologies to trace Haller's shifts in position and his reactions to others.

Obviously I have profited from Shirley Roe's solution to this problem in *Matter, Life, and Generation: Eighteenth-Century Embryology and the Haller-Wolff Debate* (Cambridge: Cambridge University Press, 1981). For Haller's generational theory, the following are essential: *Réflexions sur le système de la génération de M. de Buffon* (Geneva: Barillot et Fils, 1751); Foreword to Buffon's *Allgemeine Historie der Natur*, vol. 2 (Hamburg: Grund und Holle, 1752); *Sur la formation du coeur dans le poulet . . .*, 2 vols. (Lausanne: Bousquet, 1758); "Review of C. F. Wolff, *Theoria generationis*," *GA* 143: 1226–31; *Elementa physiologiae corporis humani*, vol. 8 (Lausanne, Bern: Bousquet, d'Arnay, Grasset; Societas Typographica, 1766).

These works must be supplemented by Haller's correspondence with Bonnet and Wolff, with statements in his edition of Boerhaave's lectures, and with his own *Primae lineae physiologiae* (Göttingen; Vandenhoek, 1747). Haller's thought on organic distribution is less complex, but no one would call it simple. The *Enumeratio methodica stirpium Helvetiae indigenarum* (Göttingen: Vandenhoek, 1742) is a preliminary study for the monumental *Historia stirpium indigenarum Helvetiae inchoata*, 2 vols. (Bern: Societatis Typographicae, 1768). A welcome addition to Haller studies in recent years is Otto Sonntag's edition of the Haller-Bonnet correspondence, *The Correspondence Between Albrecht von Haller and Charles Bonnet* (Bern: Huber, 1983). This volume traces the growth and change of Haller's opinions from week to week. The letters begin with Haller's return to Bern from Göttingen and end with his death.

Linnaean studies have profited greatly from J.M. Hulth's *Bibliographia Linnaeana: Matériaux pour servir à une bibliographie Linnéenne* (Uppsala: Almquist & Wiksell, 1907). Hulth lists Linnaeus's production year by year, disentangling most of the basic problems of chronology. Linnaean plant taxonomy is best followed in successive editions of *Genera plantarum* and *Species plantarum*, W. T. Stearn is a reliable guide to Linnaean taxonomy, and his facsimile editions of the first edition of the *Species plantarum* and the fifth edition of the *Genera plantarum* are actually preferable to the originals. Linnaeus's last thoughts on a natural method for plants are found in *Praelectiones in ordines naturales plantarum*, ed. Paul Dietrich Giseke (Hamburg: Hoffmann, 1792).

Shifts in Linnaean zoology must be traced through the editions of the *Systema naturae*. The tenth and twelfth editions are particularly important: *Systema naturae*, 10th ed., 2 vols. (Stockholm: Salvius, 1758–59); *Systema naturae*, 12th ed., 3 vols. (Stockholm: Salvius, 1766–68). The many dissertations at Uppsala with Linnaeus as *praeses* are an invaluable source of information on an astonishing number of subjects. Schreber's edition is preferable to earlier collections: *Amoenitates academicae*, 10 vols., ed. J. C. D. Schreber (Erlangen: Palm, 1785–89). Readers of Swedish will find that the selections published by the Stockholm academy contain almost everything of importance: *Valda smärre skrifter af allmänt naturvetenskapligt innehåll* and *Valda smärre skrifter af botaniskt innehåll*, vols. 2 and 4 of the *Skrifter*. The Swedish Linnaean Society publishes a yearbook that follows the progress of Linnaean research faithfully: *Svenska Linné-Sällskapets årsskrift*.

Investigation of the generation of 1760 confronts a much more difficult situation. The basic tools of research—bibliographies, scholarly editions, and biographies—are either rudimentary or lacking entirely. There are exceptions, of course; exemplary figures like Joseph Gottlieb Koelreuter and Caspar Friedrich Wolff have attracted extended commentary. Others—Peter Simon Pallas, for example, or Joseph Gaertner—are remembered for one or two important works. And some pivotal figures like Gottlieb Conrad Christian Storr have been forgotten completely. In this situation, the historian is driven to investigate the methods and projects of the sciences themselves to show how naturalists and physiologists approached and handled their fundamental problems and how they conceived the general purport of what they were doing.

Following Kant's lead, I have used two Linnaean dissertations to stand for orthodox naturalists' conceptions of natural order, *Oeconomia naturae* (1749) and *Politia naturae* (1760), both contained in the *Skrifter* (1906). The scope of the metaphysical conception of natural order should be studied, however, in Christian Wolff's *Vernünfftige Gedancken von Gott, der Welt, und der Seele des Menschen, und allen Dingen überhaupt* (Halle: Renger, 1724). The pursuit of a natural method for plants can be followed in a few texts: Joseph Pitton de Tournefort, "Isagoge in rem herbariam," *Institutiones rei herbariae*, 3d ed. (Paris: Typographia Regia, 1719); Linnaeus, "Fragmenta methodi naturalis," *Classes plantarum* in the *Skrifter* (1907); Michel Adanson, "Préface istorike sur l'état ancien et actuel de la botanike, et une théorie de cette science," *Familles des plantes*, pt. 1 (Paris: Vincent, 1763); Joseph Gaertner, *De fructibus et seminibus plantarum* (Stuttgart: Typis Academiae Carolinae, 1788); and Antoine-Laurent de Jussieu, *Genera plantarum secundum ordines naturales disposita* (Paris: Herissant, 1789).

The study of the natural method for animals in the later eighteenth century is more difficult, since each group and subgroup must be studied separately. For some groups, the literature is enormous. Linnaeus's *Systema naturae*, 10th ed., vol. 1 (Stockholm: Salvius, 1758), offers a point of departure. Most often, the *terminus ad quem* will be found in one of the works of Georges Cuvier. Between these *termini* the investigator will turn over texts, manuals, and treatises in no special order. There are no general rules, but J. Victor Carus and Henri Daudin are reliable guides. Underlying assumptions about formal relations were crystalized in the concepts of *continuity, affinity, and analogy*; shifts in the meaning of these terms are a rewarding study. Some crucial texts are Peter Simon Pallas, "De zoophytorum intermedia natura et inventione," *Elenchus zoophytorum* (The Hague: van Cleef, 1766); Johann Friedrich Blumenbach, *Beyträge zur Naturgeschichte* (Göttingen: Dieterich, 1790), pp. 111, 112; Immanuel Kant, *Erste Einleitung, GS*, sec. 3, vol. 7, p. 210; Kant, *Kritik der Urtheilskraft, GS*, sec. 1, vol. 5, p. 182. August Thienemann's "Die Stufenfolge der Dinge, der Versuch eines natürlichen Systems der Naturkörper aus dem achtzehnten Jahrhundert," *ZA* 3 (1910): 185–274, is not only suggestive but comprehensive.

Eighteenth-century naturalists did not consider questions of variation and specific diversity an integrated set of problems, and the different kinds of deviation from natural order must be studied separately. Here are some important texts. The orthodox view of species and variation will be found in Linnaeus's *Fundamenta botanica* (Amsterdam: Schouten, 1736) or the *Critica botanica* (Leiden: Wishoff, 1737). Buffon first stated his alternative in the "Premier Discours," in vol. 1 of the *Histoire naturelle*. The problem of variation in plants in the eighteenth century is succinctly analyzed in Augustin-Pyrame de Candolle, *Théorie élémentaire de la botanique*, 3d ed. (Paris: Roret, 1844).

Opinions about plant hybridization can be studied in Linnaeus's dissertation, "De sexu plantarum," *Skrifter* (1908), and in Joseph Gottlieb Koelreuter's *Vorläufige Nachricht von einigen das Geschlecht der Pflanzen betreffenden Versuchen und Beobachtungen nebst Fortsetzungen 1, 2, und 3 (1761–1766)*, ed. W. Pfeffer (Leipzig: Engelmann, 1893). An equivalent study for animals is Peter Simon Pallas's "Mémoire sur la variation des animaux, " *Acta*, 1780, *pars posterior* (1784): 69–102. Immanuel Kant's precritical utterance on the human race (1775), "Von den verschiedenen Racen der Menschen," *GS*, sec. 1, vol. 2, pp. 427–43, should be supplemented with two later papers, "Bestimmung des Begriffs einer Menschenrasse" (1785) and "Über den Gebrauch teleologischer Prinzipien in der Philosophie" (1788), in *GS*, sec. 1, vol. 8, pp. 91–106, 157–84.

Johann Friedrich Blumenbach's early statement on the human race, "De generis humani varietate nativa," appears in *The Anthropological Treatises*, ed. and trans. Thomas Bendyshe (London: Anthropological Society, 1865), 65–143. Racial questions were part of Blumenbach's ambitious synthesis of orthodox taxonomy, the theory of generation, and the philosophy of Kant. For the effect on the next generation, see Christoph Girtanner, *Ueber das kantische Prinzip für die Naturgeschichte* (Göttingen: Vandenhoek und Ruprecht, 1796), and J. K. W. Illiger, *Versuch einer systematischen vollständigen Terminologie für die Thierreich und das Pflanzenreich* (Helmstädt: Fleckeisen, 1800).

Geographical history did not spring fully formed from the head of Eberhard Zimmermann. Both Linnaeus and Buffon tried to articulate the problems of geographical distribution theoretically: Linnaeus, "De telluris habitabilis incremento," in *Skrifter* (1906); Buffon, "Cinquième Époque," *Oeuvres* (Paris: Presses Universitaires de France, 1954), pp. 168–76. Inconsistencies in these theories led to Zimmermann's *Geographische Geschichte*, 3 vols. (Leipzig: Weygand, 1778–83). Zimmermann also had at his disposal an immense number of facts contained in travelers' accounts, institutional collections, faunas, and floras. Here are three very important works: Johann Georg Gmelin, *Flora sibirica*, 4 vols. (St. Petersburg: Typographiae Academiae Scientarum, 1742–1769); Peter Simon Pallas, *Reise durch verschiedene Provinzen des russischen Reichs*, 3 vols. (Graz: Akademische Druck und Verlagsanstalt, 1967); Johann Reinhold Forster, *Observations Made During a Voyage Round the World* (London: Robinson, 1778). The list could be extended indefinitely. Two later summations should be consulted: Augustin-Pyrame de Candolle, "Géographie botanique," in *Dictionnaire des sciences naturelles*, vol. 10 (Strasbourg: Levrault and Le Normant, 1820); Nils von Hofsten, "Zur älteren Geschichte des Diskontinuitätsproblems," *ZA* 7 (1919): 197–353.

In the science of generation, the important proponents of mechanistic epigenesis are Buffon, Pierre-Louis Maupertuis, and Caspar Friedrich Wolff. The essential works are: Buffon, "Histoire générale des animaux," *Oeuvres*, pp. 233–89; Maupertuis, *Vénus physique* (Paris: Montaigne, 1980); Caspar Friedrich Wolff, *Theoria generationis* (Halle: Hendel, 1759); Wolff, *Theorie von der Generation* (Berlin: Birnstiel, 1764); Wolff, *De formatione intestinorum*, pts. 1 and 2, *NC* 12 (1766–67): 403–507; Wolff, *De formatione intestinorum*, pt. 3, *NC* 13 (1768): 478–530. Haller's defense of *evolutio* is in *Réflexions sur le systeme de la génération de M. de Buffon*, mentioned earlier. Charles Bonnet's position is found in Otto Sonntag, ed., *The Correspondence between Albrecht von Haller and Charles Bonnet* (Bern: Huber, 1983). Wolff's confrontation with Blumenbach is in Wolff's *Von der eigenthümlichen und wesentlichen Kraft* (St. Petersburg: Kayserlichen Academie der Wissenschaften, 1789). Blumenbach's version of epigenesis is in *Über den Bildungstrieb und das Zeugungsgeschäfte* (Göttingen: Dieterich, 1781). His contribution to Wolff's prize competition is the *Erste Abhandlung über die Nutritionskraft* (St. Petersburg: Kayserlichen Academie der Wissenschaften, 1789).

The study of Kant's critique of teleological judgment is an analysis of two texts, the *Erste Einleitung* in *GS*, sec. 3, vol. 7, pp. 193–251, and the *Kritik der Urtheilskraft*, *GS*, sec. 1, vol. 5, pp. 167–485. Untangling Kant's subtleties is never easy; Konrad Marc Wogau's *Vier Studien zu Kants "Kritik der Urtheilskraft*," Uppsala universitets årsskrift 2 (Uppsala: Lundequistska, 1938) is refreshingly stringent and scrupulously fair.

This bibliographical afterword is no more than a large-scale map of largely unknown territory in scientific history. More work urgently needs to be done on the later eighteenth century. New and important discoveries are not only possible but likely. A couple of cautionary remarks should be added. Many of the tools of scientific history are wanting and will have to be created; it will be many years before the history of life sciences for this period attains the standard set by the history of the exact sciences. A reading knowledge of Latin, German, and French remains an absolute and unavoidable necessity.

Index

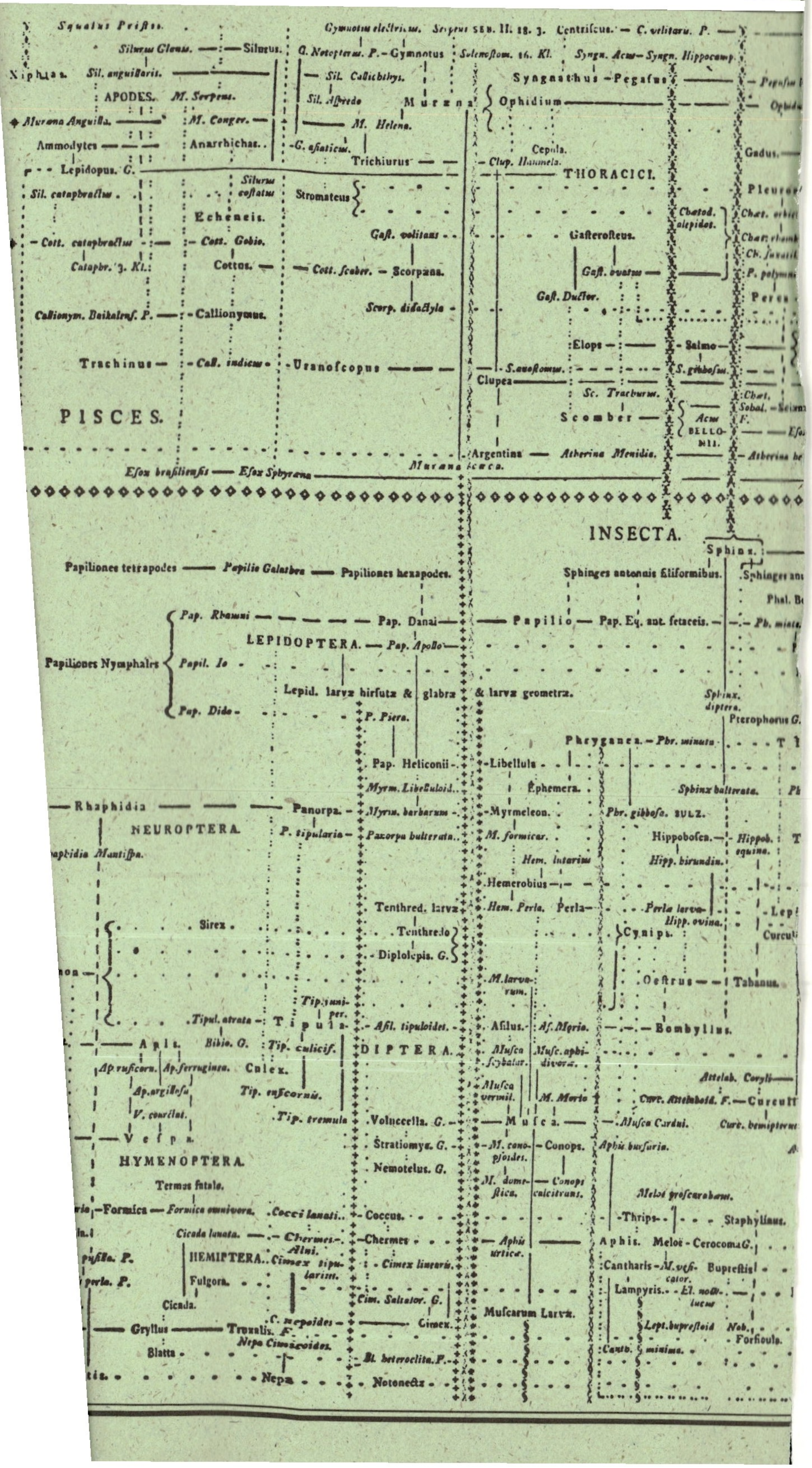
PISCES
THORACICI
INSECTA
LEPIDOPTERA
NEUROPTERA
DIPTERA
HYMENOPTERA
HEMIPTERA

Library of Congress Cataloging-in-Publication Data

Larson, James L.
Interpreting nature : the science of living form from Linnaeus to Kant / by James L. Larson.
p. cm.
Includes bibliographical references (p.) and index.
ISBN 0-8018-4840-7 (hc : acid-free paper)
1. Natural history—Europe, Northern—History—18th century.
2. Physiology—Europe, Northern—History—18th century. I. Title.
QH21.E853L37 1994
574′.094′09033—dc20 94-2904